THE SOLAR SYSTEM

THE MOON
AND OTHER
SMALL BODIES

THE SOLAR SYSTEM

THE MOON AND OTHER SMALL BODIES

SALEM PRESS
A Division of EBSCO Publishing

Ipswich, Massachusetts

GREY HOUSE PUBLISHING

CONTENTS

CONTRIBUTORS

Stephen R. Addison

Michael S. Ameigh

Raymond D. Benge, Jr.

John L. Berkley

Larry M. Browning

Dennis Chamberland

Joseph Dewey

Richard R. Erickson

David G. Fisher

George J. Flynn

Sarah J. Greenwald

Alexander A. Gurshtein

Brian Jones

Karen N. Kähler

Firman D. King

Richard S. Knapp

James C. LoPresto

Caryn E. Neumann

Steven C. Okulewicz

Robert J. Paradowski

Charles W. Rogers

Lawrence H. Shirley

Paul P. Sipiera

Billy R. Smith, Jr.

Joseph L. Spradley

Jill E. Thomley

James L. Whitford-Stark

APOLLO PROGRAM

Category: Space Exploration and Flight

The Apollo Program, designed to ensure America's international leadership in space exploration, resulted in the first landing of humans on the Moon. Apollo astronauts performed experiments and returned rock samples to Earth that helped determine the age and origin of the Moon.

Overview

In 1960, planners at the National Aeronautics and Space Administration (NASA) selected a crewed lunar landing as the follow-up to the Mercury effort to place a man in Earth orbit. In December, 1960, just before leaving office, President Dwight D. Eisenhower advised NASA officials that he would not approve the lunar landing project. However, after the Soviets sent Yuri A. Gagarin into Earth orbit in April, 1961, the new U.S. president, John F. Kennedy, announced on May 25, 1961, the plan "before this decade is out, of landing a man on the Moon and returning him safely to the earth." The project to accomplish Kennedy's objective was named Apollo.

The Apollo Spacecraft

To launch Apollo, NASA designed and built a huge, three-stage rocket, the Saturn V, which stood 363 feet tall and developed 7.5 million pounds of thrust at liftoff. In order to minimize the weight of the spacecraft, the Apollo engineers planned a lunar orbit rendezvous technique, requiring the Apollo spacecraft to have a modular design, consisting of three separate units, the Command Module, the Service Module, and the Lunar Module.

The Command Module, built by North American Rockwell, served as the control center for the spacecraft and provided 210 cubic feet of living and working space for the astronauts. It was designed to carry three astronauts from the earth to an orbit around the Moon and back. It was shaped like a cone, with a height of 10 feet 7 inches, a maximum diameter of 12 feet 10 inches, and an approximate weight of 13,000 pounds. The Command Module was pressurized, so the astronauts could live and work without wearing spacesuits. The wide end of the cone was a blunt heatshield, covered with layers of special ablative material designed to burn away during reentry, dissipating the extreme heat of atmospheric friction.

The cylindrical Service Module, built by North American Rockwell, had a diameter of 12 feet and 10 inches and a length of 22 feet and 7 inches. It carried the electrical power systems, most of the electronics, and the life support gases. It also carried the computer system for guidance and navigation, the communications transmitters and receivers, and the oxygen and hydrogen used by the life-support and energy-generation systems. The Service Module's rocket engine produced 22,000 pounds of thrust. This rocket engine was used to slow the spacecraft to enter lunar orbit and then to speed it up for the return to Earth. Fully fueled, the Service Module weighed about 53,000 pounds.

The Lunar Module, built by the Grumman Aircraft Engineering Corporation, was designed to detach from the Command and Service Modules while they orbited the Moon and to carry two astronauts down to the lunar surface. The Lunar Module was a two-stage rocket, with each stage carrying its own fuel supply. The lower stage carried a 9,700-pound-thrust rocket engine to slow down the Lunar Module for a gentle touchdown on the lunar surface. Four landing legs, each with a landing pad to distribute the weight of the spacecraft over a larger area of the lunar soil, were attached to the Lunar Module descent stage. One of the landing legs was equipped with a ladder to allow the astronauts to climb down to the lunar surface. The upper stage of the Lunar Module consisted of a pressurized compartment providing life support for the two-man crew and an ascent engine to return the crew compartment to lunar orbit. The lower stage of the Lunar Module served as a launching pad for the upper stage. With the landing legs extended, the Lunar Module was 22 feet and 11 inches tall and weighed about 32,000 pounds.

The Apollo Flights

On January 27, 1967, during a preflight test, a fire swept rapidly through the Apollo Command Module. The three astronauts participating in the test, Roger Chaffee, Virgil "Gus" Grissom, and Edward White, were killed in the fire. After the fire, NASA officials designated the test as Apollo 1, honoring the crew. An extensive investigation of the fire showed numerous design flaws in the Apollo Command Module, and crewed launchings were postponed for more than a year while an extensive redesign was conducted.

Apollo 7, the first manned test of the Command and Service Modules, was launched from Cape Kennedy, Florida, on October 11, 1968, on a Saturn IB rocket. Apollo 7 was the only crewed Apollo mission launched on a Saturn IB rocket, which was powerful enough to carry the Command Module and the Service Module into Earth orbit, but could not lift the full Apollo assembly, including the Lunar Module. The spacecraft crew consisted of commander Walter M. Schirra, Jr., Donn F. Eisele, and Walter Cunningham, who held the title of Lunar Module pilot despite the lack of a Lunar Module on the Apollo 7 mission. The crew orbited the earth 163 times and spent almost eleven days in space, demonstrating the reliability of the Command and Service Modules for a time comparable to that of a round trip to the Moon.

Apollo 8, launched on December 21, 1968, was the first crewed mission using the Saturn V rocket, and the first mission to take humans to the Moon and back. The three-person crew consisted of Frank Borman, the commander; James A. Lovell, Jr., the Command Module pilot; and William A. Anders, the Lunar Module pilot. Apollo 8 tested the flight path and operations for the trip to the Moon and back and demonstrated that the Apollo Command Module could successfully reenter the earth's atmosphere at the high speed of a return from the Moon.

The Apollo 9 mission, launched on March 3, 1969, was the first crewed flight employing all three components of the Apollo spacecraft. The crew, consisting of astronauts James A. McDivitt, the commander; David R. Scott,Scott, the Command Module pilot; and Russell L. Schweickart, the Lunar Module pilot, made 152 orbits of the earth. They demonstrated the crew transfer procedures and the rendezvous and docking procedures between the Command Module and the Lunar Module.

The final test of the Apollo spacecraft came with Apollo 10, launched on May 18, 1969. Apollo 10 was a complete Apollo lunar landing mission without an actual landing on the Moon. On the fifth day of the mission, astronauts Thomas Stafford, the commander, and Eugene Cernan, the Lunar Module pilot, descended inside the Lunar Module to within 14 kilometers of the lunar surface, while John W. Young remained in lunar orbit in the Command Module.

The Saturn V rocket carrying Apollo 11 lifted off from the NASA John F. Kennedy Space Center in Florida on July 16, 1969. The Command Module, named *Columbia* carried astronauts Neil Armstrong, the commander; Edwin "Buzz" Aldrin, the Lunar Module pilot; and Michael Collins, the Command Module pilot, to the Moon and back. Astronauts Armstrong and Aldrin landed on the Moon in the Lunar Module, named *Eagle*, on July 20, 1969. Michael Collins remained alone in the *Columbia*, orbiting the Moon. *Columbia* served as a communications link between the astronauts on the Moon and mission control in Houston, Texas. After 28 hours on the Moon, the upper stage of the Lunar Module carried Armstrong and Aldrin back into orbit around the Moon, where they rendezvoused and docked with the *Columbia*. *Columbia*, the only part of the spacecraft to return to Earth, landed in the Pacific Ocean on July 24, 1969.

Apollo 12, launched on November 14, 1969, carried astronauts Charles Conrad, Jr., the commander; Richard F. Gordon, the Command Module pilot; and Alan L. Bean, the Lunar Module pilot. Astronauts Conrad and

The Apollo Programhad to devise spacecraft that not only would take humans out of Earth's atmosphere, but also land them on the Moon. The Lunar Module transported astronauts from their primary craft to the Moon's surface. (NASA CORE/Lorain Valley JVS)

Bean landed on the Moon in the Sea of Storms, less than 600 feet from the site where the Surveyor 3 spacecraft had landed on April 20, 1967. The astronauts recovered pieces from the Surveyor 3 to allow scientists to assess the effects of the craft's two-year exposure to the lunar environment. They also collected 75 pounds of rocks and soil for return to Earth and deployed the Apollo Lunar Surface Experiment Package to perform scientific experiments on the Moon.

Apollo 13, carrying astronauts James A. Lovell, Jr., the commander; John L. Swigert, Jr., the Command Module pilot; and Fred W. Haise, Jr., the Lunar Module pilot, lifted off on April 11, 1970. About 56 hours into the flight, an explosion in one of the oxygen tanks in the Service Module crippled the spacecraft. The crew was forced to orbit the Moon and return to the Earth without landing. The astronauts spent much of the flight in the Lunar Module, using its oxygen and electrical supplies, because of the damage to the Service Module. The astronauts landed safely on Earth on April 17, 1970.

Apollo 14, carrying astronauts Alan B. Shepard, the commander; Stuart A. Roosa, the Command Module pilot; and Edgar D. Mitchell, the Lunar Module pilot, was launched on January 31, 1971. Astronauts Shepard and Mitchell landed on February 5, 1971, within 160 feet of the target point, in the Fra Mauro region of the Moon, the intended landing site of the Apollo 13 mission. During a 4-hour, 20-minute period of extravehicular activity, Shepard and Mitchell climbed up the side of Cone Crater, providing the first experience of climbing and working in hilly terrain in the bulky spacesuits. The astronauts collected 94 pounds of lunar soil and rocks. The upper stage of the Lunar Module lifted off from the lunar surface on February 6, 1971, after 33.5 hours on the Moon. After the crew transferred to the Command Module, the Lunar Module ascent stage was guided to impact on the lunar surface, producing a seismic signal that was recorded by instruments deployed on the lunar surface by Apollo 12 and Apollo 14. The Command Module landed in the Pacific Ocean on February 9, 1971.

By 1970, public interest in lunar exploration had waned and federal budget cuts forced NASA to sacrifice current projects in order to support future ones. Apollo 15, 16, and 17 would be equipped to travel farther, stay longer, and perform more experiments than had previous missions.

Apollo 15, carrying astronauts David R. Scott, the commander; Alfred J. Worden, the Command Module pilot; and James B. Irwin, the Lunar Module pilot, was launched on July 26, 1971. Apollo 15 was the first in a series of advanced missions, carrying the Lunar Rover (LRV), which astronauts Scott and Irwin used to explore the Hadley Rille region of the Moon. The LRV allowed astronauts to travel tens of kilometers from the Lunar Module, in contrast to the hundreds of meters traveled in previous missions. The astronauts collected 173 pounds of samples from the low lunar plains, the Apennine Mountains, and the Hadley Rille, a long, narrow, winding valley. Although Apollo 15's atmospheric entry was normal, one of the three parachutes that slowed the Command Module's descent collapsed before landing. Nonetheless, the Command Module landed safely on August 7, 1971.

Apollo 16, carrying astronauts John W. Young, the commander; Thomas K. Mattingly II, the Command Module pilot; and Charles M. Duke, Jr., the Lunar Module pilot, was launched on April 16, 1972. This mission landed in the Descartes region, where astronauts Young and Duke collected 209 pounds of soil and rocks and used an ultraviolet camera and spectrograph to perform the first astronomical measurements from the surface of the Moon. The Apollo 16 crew returned to Earth on April 27, 1972.

Apollo 17, carrying astronauts Eugene A. Cernan, the commander; Ronald E. Evans, the Command Module pilot; and Harrison H. Schmitt, the Lunar Module pilot and, as a trained geologist, only scientist to visit the Moon, was launched on December 7, 1972. On this mission, astronauts Cernan and Schmitt conducted the longest LRV traverse on a single extravehicular activity, a trip of about 100 kilometers. They collected the largest amount of lunar soil and rock ever returned to Earth. Apollo 17's return to Earth on December 19, 1972, marked the end of U.S. efforts to send humans to the Moon.

Results of the Apollo Program

The major objective of the Apollo Program was accomplished with the landing of twelve American astronauts on the Moon and their safe return to Earth. These landings demonstrated the capability of American engineering, restoring American prestige by finally beating the Soviet Union in the space race. Scientists studying lunar rock samples were finally able to determine the age and origin of the Moon, finding that the Moon formed about 4,560,000,000 years ago, probably from the debris ejected when an asteroid struck Earth.

George J. Flynn

Further Reading

Brooks, Courtney G., James M. Grimwood, and Lloyd S. Swenson. *Chariots for Apollo: A History of Manned Lunar Spacecraft.* Washington, D.C.: National Aeronautics and Space Administration, 1979. NASA's official history of the Apollo Program, focusing on the design, construction, and flight of the Apollo spacecraft.

Chaikin, Andrew L. *A Man on the Moon: The Voyages of the Apollo Astronauts.* New York: Viking Press, 1994. An extensive historical account of the Apollo Program, beginning with the Apollo 1 fire and continuing through the successful Moon landing.

Logsdon, John W. *The Decision to Go to the Moon.* Cambridge, Mass.: MIT Press, 1970. An extensive history of the Apollo Program, focusing on the decisions faced by political, industrial, and NASA officials that shaped the Apollo spacecraft and the lunar landing program. Includes a comprehensive bibliography.

ASTEROIDS

Category: Small Bodies

Asteroids are minor bodies of a wide variety of sizes in orbit around the Sun, primarily but not exclusively located between the orbits of Mars and Jupiter. They provide important clues regarding the early history of the solar system, including the effect of their collisions on the surfaces of planets or their satellites. A class popularly referred to as near-Earth asteroids threaten to impact Earth.

Overview

Although discovery of the first asteroid was accidental, it came as no surprise to the astronomical community of the day. In 1766, German astronomer Johann Titius (1729-1796) observed that the positions of the planets could be approximated very closely by a simple empirical rule. Adding 4 to each number in the sequence {0, 3, 6, 12, 24, 48 . . . } and dividing the sum by 10 yields the mean planetary distances from the Sun in astronomical units (the distance from the Earth to the Sun is one astronomical unit, or 1 AU). The exception to this rule is the fifth element in that purely mathematical sequence, where an apparent gap occurs at 2.8 AU. It must be noted that this is just an empirical observation with no known physical basis. This rule was publicized by Johann Bode (1747-1826) and led to a search for a missing planet in the gap between Mars at 1.5 AU and Jupiter at 5.2 AU.

On January 1, 1801, the Sicilian astronomer-monk Giuseppe Piazzi (1746-1826) accidentally discovered a moving object during a routine star survey. He named it Ceres, for the patron goddess of Sicily. Soon its orbit was calculated by Carl Friedrich Gauss (1777-1855). At 2.77 AU, Ceres was found, coincidentally, to conform closely to the Titius-Bode rule. However, since Ceres seemed to be too small to be classified as a planet, the search continued.

In March, 1802, German astronomer Heinrich Olbers (1758-1840) found a second minor body at the same predicted distance. He named it Pallas. In 1803, Olbers proposed that meteorites come from an exploded planet near 2.8 AU. This possibility led to a continued search resulting in the discovery of Juno in 1804 and Vesta in 1807. The latter discovery again was made by Olbers. It took quite some time for a fifth small body to be discovered (in 1845), but by 1890 the total had reached three hundred. These bodies came to be called "asteroids" for their faint, starlike images. In 1891, German astronomer Max Wolf (1863-1932) began using a long-exposure camera to detect asteroids. Since then, thousands of asteroids have been registered in the official catalog of the Institute of Theoretical Astronomy in Leningrad.

Asteroids are usually referred to officially by both a number and a name, such as 3 Juno or 1,000 Piazzi. About one hundred newly numbered asteroids are cataloged each year. Sky surveys indicate as many as 500,000 asteroids that are large enough to appear on telescopic photographs. Most asteroids are found within the main asteroid belt, which extends from 2.1 to 3.4 AU, and about half are between 2.75 and 2.85 AU. Asteroids revolve around the Sun in the same direction as the planets but tend to have more elliptical orbits. Their orbits are inclined up to 30° to the ecliptic plane, but they are far less eccentric than comet orbits. The smallest asteroids are a few kilometers wide; the largest, 1 Ceres (now considered a "dwarf planet"), about 1,000 kilometers wide. In 1867, American astronomer Daniel Kirkwood (1814-1895) discovered gaps in the asteroid belt where relatively few asteroids are found. These so-called Kirkwood gaps occur where asteroids have orbital periods that are simple fractions of the twelve-year revolution period of the giant planet Jupiter about the Sun, resulting in periodic gravitational influences called resonances. Such depletions occur, for example, at about 3.3 AU (where the periods have a

An artist's rendering of one of two asteroid belts of Epsilon Eridani, a solar system located in the constellation Eridanus. (NASA/JPL-Caltech)

six-year, 1:2 resonance with Jupiter) and 2.5 AU (a four-year, 1:3 resonance); other resonances, however, act to stabilize certain asteroids, such as the Hilda group at 4 AU (2:3 resonance), which is named for 153 Hilda.

Some asteroids have orbits departing greatly from the main belt. In 1772, French mathematician Joseph Lagrange (1736-1813) showed that points in Jupiter's orbit 60° ahead of and behind the planet are gravitationally stable (1:1 resonance). In 1906, Max Wolf discovered the first so-called Trojan asteroid, 588 Achilles, at the Lagrangian point 60° ahead of Jupiter. Subsequent discoveries have revealed several hundred Trojan asteroids. Those ahead of Jupiter are named for Greek heroes, and those behind are named for Trojan heroes; there is one Greek spy (617 Patroclus asteroid) in the Trojan group, and one Trojan spy (624 Hektor) in the Greek group. Hektor is the largest known Trojan asteroid, at about 150 by 300 kilometers, and is the most elongated of the more massive asteroids. At least two objects have orbits that extend beyond Jupiter: 944 Hidalgo, which may be a burned-out cometary nucleus, and 2060 Chiron, whose orbit extends beyond Saturn.

Some asteroids depart from the main belt over only part of their orbit. Mars-crossing Amor group bodies have elongated orbits that carry them inside Mars's orbit but still keep them well outside Earth's orbit. The Martian satellites Phobos and Deimos have long been suspected by many to be captured asteroids, perhaps from this group. Apollo group members come inside Earth's orbit. (The groups were named for their first examples, discovered in 1932.) Estimates indicate about thirteen hundred Apollos ranging in from 0.4 to 10 kilometers across, with an estimated average Earth-collision rate of about one in 250,000 years. The closest known approaches were Hermes, in 1937, at about 780,000 kilometers, and 1566 Icarus, in 1968, at about 6 million kilometers. Smaller Apollos may be an important source of meteorites, and 100-meter objects capable of making a 1-kilometer crater strike Earth about every two thousand years. Aten-type asteroids are Earth-crossers with elliptical orbits smaller than Earth's. Some asteroids appear to be grouped in families that may be the fragments resulting from an earlier collision between asteroids.

Chemical and physical characteristics of asteroids are mostly determined by remote-sensing techniques that study electromagnetic radiation reflected off their surfaces. More than five hundred asteroids have been studied by remote sensing and radar astronomical techniques.

These studies have indicated asteroidal compositions similar to those of meteorites. Comparison with reflected light from meteorites suggests several classes. Rare E-type asteroids possess the highest albedo (23 to 45 percent reflection). They appear to be related to enstatite (a magnesium silicate mineral) chondrites, and are concentrated near the inner edge of the main belt. About 10 percent of asteroids are S-type; they have relatively high albedos (7 to 23 percent) and appear reddish in color. They likely are related to stony chondrites, are found in the inner to central regions of the main belt, and generally range in size from 100 to 200 kilometers. The largest S-type is 3 Juno, at about 250 kilometers in diameter. Much smaller Apollo asteroids are also in this category. A few asteroids in the middle belt are classified as M-type, since their reflected light (7 to 20 percent) reveals evidence of large amounts of nickel-iron metals on their surface, similar to iron or stony-iron meteorites.

About three-quarters of all asteroids are C-type, having relatively low albedos (2 to 7 percent) and grayish colors similar to that of the Earth's moon. They are found in the outer belt and among the Trojans. They resemble carbonaceous chondrite meteorites, containing water-bearing silicate-based and carbon-based

Facts About Selected Asteroids

Name	Diameter (km) or Dimensions	Mass (10^{15}kg)	Rotational Period (hrs)	Orbital, Period (yrs)	Distance from Sun (AU)
Ceres (dwarf planet)	960 × 932	870,000	9.075	4.60	2.767
Palas	570 × 525 × 482	318,000	7.811	4.61	2.774
Juno	240	20,000	7.210	4.36	2.669
Vesta	530	300,000	5.342	3.63	2.362
Eugenia	214	6,100	5.699	4.49	2.721
Siwa	110	1,500	18.5	4.52	2.734
Ida	58 × 23	100	4.633	4.84	2.861
Mathilde	66 × 48 × 46	103.3	417.7	4.31	2.646
Eros	33 × 13 × 13	7.2	5.270	1.76	1.458
Gaspra	19 × 12 × 11	10	7.042	3.29	2.209
Icarus	1.4	0.001	2.273	1.12	1.078
Geographos	2.0	0.004	5.222	1.39	1.245
Apollo	1.6	0.002	3.063	1.81	1.471
Chiron	180	4000	5.9	50.70	13.633
Shipka	—	—	—	5.25	3.019
Rodari	—	—	—	3.25	2.194
McAuliffe	2-5	—	—	2.57	1.879
Mimistrobell	—	—	—	3.38	2.249
Toutatis	4.6 × 2.4 × 1.9	0.05	130	3.98	2.512
Nereus	2	—	—	1.82	1.490
Castalia	1.8 × 0.8	0.0005	—	1.10	1.063
Otawara	5.5	0.2	—	3.19	2.168
Braille	2.2 × 1.0	—	—	3.58	2.341

Source: Data are from the National Aeronautics and Space Administration/Goddard Space Flight Center, National Space Science Data Center.

minerals along with some organic compounds (about 1 percent). The largest of all the asteroids, 1 Ceres (now considered a dwarf planet), is in this category. Some evidence supports the claim that Ceres has a mixture of ice and carbonaceous minerals on its surface. Dark reddish, D-type asteroids are found in the same regions and have similar albedos.

About 10 percent of asteroids remain unclassified and are designated as U-type. In general, asteroids with low-temperature volatile materials lie farther from the Sun, whereas those in the inner part of the main belt are richer in high-temperature minerals, displaying little evidence of volatile water and carbon compounds.

Many asteroids exhibit periodic variations in brightness that suggest irregular shapes and rotation. Their measured rotational periods range from about three to thirty hours. There is some evidence that S-type asteroids rotate faster than C-type asteroids but more slowly than M-type asteroids. Large asteroids (greater than 120 kilometers) rotate more slowly with increasing size, but small asteroids rotate more slowly with decreasing size, suggesting that large asteroids may be primordial bodies, while smaller ones may be fragments produced by collisions. Calculations show that rotation rates longer than two hours produce centripetal forces weaker than gravity, which indicates that loose debris can exist on the surface of even the fastest known rotating asteroid, the Apollo object 1566 Icarus, which has a 2.25-hour rotation rate. Polarization studies of light reflected from asteroids indicate that many do have dusty surfaces.

Named after the Greek god of love, Eros is an S-type asteroid belonging to the Amor group. As the second-largest of the near-Earth asteroids, it is larger than the asteroid generally accepted to have been responsible for the extinction of the dinosaurs that impacted Earth near the Yucatán peninsula 65 million years ago. Eros is 13 by 13 by 33 kilometers in size. It was the first asteroid recognized to approach inside the orbit of Mars, and in 1975 it became the first asteroid to be studied with Earth-based radars.

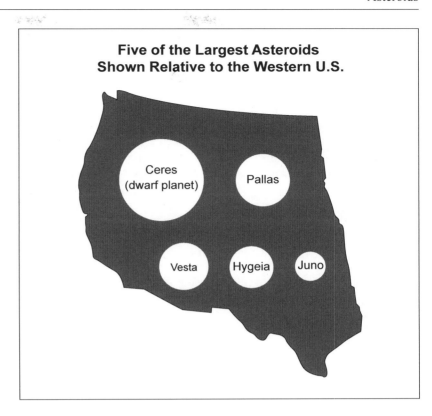

Five of the Largest Asteroids Shown Relative to the Western U.S.

Ceres (dwarf planet)

Pallas

Vesta   Hygeia   Juno

Computer models suggest the possibility that larger asteroids have a deep layer of dust and rock fragments (or regolith) similar to that on the surface of the Moon. Asteroids with diameters larger than 100 kilometers are believed to have undergone a process of differentiation in which heavier metals sank to the core, leaving a stony surface of lighter materials later pulverized by collisions to form a layer of dust.

Asteroid elongations can be estimated from the change in brightness, which can vary by a factor of three or more. For example, radar evidence indicates that the unusual Trojan asteroid 624 Hektor (150 by 300 kilometers) may be a dumbbell-shaped double asteroid. Kilometer-scale asteroids have been observed with lengths up to six times greater than their width. Main-belt asteroids tend to be less elongated than Mars-crossers of the same size, perhaps because of greater erosion from collisions in the belt. Asteroids larger than about 400 kilometers tend to be more spherical, since their gravitational attractions exceed the strength of their rocky materials, causing deformation and plastic flow into a more symmetric shape.

An asteroid's size occasionally can be determined quite accurately by timing its passage in front of a star, that is, in a stellar occultation. In a few cases, stellar light has been occulted more than once in a single passage, indicating that asteroids may possess satellites. Radar-based studies have confirmed this theory. Also, as the Galileo spacecraft flew through the main belt on its way to enter orbit in the Jupiter system, it imaged a satellite revolving about an asteroid. The irregularly shaped asteroid Ida was discovered to have a small satellite later named Dactyl. Ida is a member of the Koronis family and of S-type, which is 56 by 24 by 21 kilometers in size and rotates once around its own axis every 278 minutes. Dactyl is only 1.2 by 1.4 by 1.6 kilometers in size and is also of S-type. This strongly suggests that it was created when a larger asteroid smashed into Ida. Previously, the Galileo spacecraft had also provided the first close-up images of an asteroid, when it passed within five thousand kilometers of the 19- by 12- by 11-kilometer-sized S-type asteroid Gaspra on October 29, 1991. Gaspra has an irregular shape, one resembling a potato.

The distribution of asteroid sizes and masses supports the idea that many have undergone a process of fragmentation. Typical relative velocities of encounter, about 5 kilometers per second in the main belt, are quite adequate to fragment most asteroids. Ceres contains nearly half the mass of all the asteroids, but it is more than three times smaller than the Moon and about fifty times less massive. About 80 percent of the total mass of all asteroids is contained in the four largest ones, and only about ten are larger than 300 kilometers. Studies suggest that the main belt was several times more massive in the past but that in the process of fragmentation, the smallest dust particles were removed by radiation pressure from the Sun.

Interest in asteroids increased when strong evidence was advanced to solve the mystery of the demise of the dinosaurs 65 million years ago. Physicist Luis Alvarez and his geologist son Walter sampled the worldwide clay layer that marks the end of the Cretaceous period and the start of the Tertiary period (the so-called K-T boundary, which essentially marks the demarcation between the age of dinosaurs and the rise of mammals within the fossil record). This thin layer of clay is enriched in the rare elements of iridium and osmium, having levels more akin to asteroids than Earthly materials. Thus, the impact theory for killing off the dinosaurs was proposed, and largely accepted except by certain portions of the paleontology community. That is, until a crater dated to 65 million years was discovered off the coast of the Yucatán peninsula. Some still insist that more than an asteroid impact was necessary to

account for the observed diminishment of dinosaur species leading up to the extinction event 65 million years ago. However, the majority of the scientific community has come to accept the asteroid impact theory, at least as the principal cause of the sudden mass extinction at the end of the Cretaceous period. Since this event marks the boundary between the Cretaceous and Tertiary periods, it is often referred to as the K-T event.

This spurred interest in asteroid and comet impacts causing extreme environmental damage to the Earth at other times in the past, along with a desire to search for near-Earth asteroids that might represent a threat in the future. Twenty-five years after the proposal that an asteroid impact killed the dinosaurs received initial lukewarm acceptance by paleontologists; some researchers proposed that an even bigger asteroid (or comet) impact was responsible for the so-called Great Dying, the mass extinction at the end of the Permian period that closed out the Paleozoic era. At the end of the Permian 248 million years ago, more than 95 percent of all species died off rather suddenly; life nearly did not make it into the Mesozoic era, during which the dinosaurs eventually arose to dominance.

Researchers point to a large crater in the Antarctic (1.5 kilometers under the ice pack that dates to the time of the Permian mass-extinction event) as well as heavily jumbled areas in Siberia (known as the Siberian Traps), that might have received tremendous seismic energy after impact energy would have undergone antipodal focusing off Earth's core. The Siberian Traps also was an area of tremendous volcanic activity at the end of the Permian period. Was this coincidental or the result of an impact with antipodal focusing of seismic energy? In 2008 this theory remained highly speculative, rather than enjoying the widespread acceptance of the K-T event that killed the dinosaurs. However, if the theory is correct, such an event underscores the danger posed by asteroid and comet impacts on Earth.

Impact of even a small asteroid could pose a tremendous threat to human civilization. Throughout the 1990's and the early twenty-first century, a number of newly discovered near-Earth asteroids were thought to have a significant chance of hitting Earth in the quite near future. However, in each of those cases, additional observations refined the asteroid's orbit to the point where it was clear it would not hit Earth after all. There remained one major exception, however. Discovery of the asteroid 99942 Apophis, a member of the Aten group, led to major concern beginning in late 2004 that this 350-meter-across rock had a relatively worrisome potential to impact Earth in 2029. Precise observations of Apophis's orbit, ranging from 0.746 AU to 1.099 AU, dramatically lowered the probability that it would strike the Earth. However, Apophis would indeed pass within the altitude of geosynchronous satellites, less than 36,000 kilometers from Earth's surface. If Apophis passed within a special corridor only 400 meters across, gravitational influences could cause it to return and strike the Earth on Friday, April 13, 2036.

The Torino scale assesses the relative impact hazard of an asteroid impact. For a time after its discovery, Apophis rated a level 4 on the Torino scale; that is the highest level of threat. Further orbit refinements lowered the threat assessment to a level 0 threat, but, after realization of the possible return in 2036, it was raised to Level 1. Although Apophis will come very close to Earth in 2029, refinement of available orbital data has since determined that

the chances of Apophis hitting the Earth in 2036 are more comforting: less than 1 in 45,000. The 2036 encounter will set up another close encounter the following year, but the chances that this would result in an Earth impact are calculated to be less than 1 in 12.3 million. Nevertheless, Apophis points out the absolute requirement for close monitoring of asteroids, particularly the near-Earth ones, and the development of means whereby asteroids could be deflected or destroyed in order to preserve Earth's biosphere and save human civilization. This sort of natural megadisaster is one of the few that humans have the potential to mitigate or prevent if action is taken sufficiently early once the threat is identified.

Hollywood has even taken notice of the asteroid or comet impact threat to Earth. Several scientifically incorrect action movies were produced, some of which were popularly received. Many of these movies, such as *Armageddon* (1998), portray the use of some type of

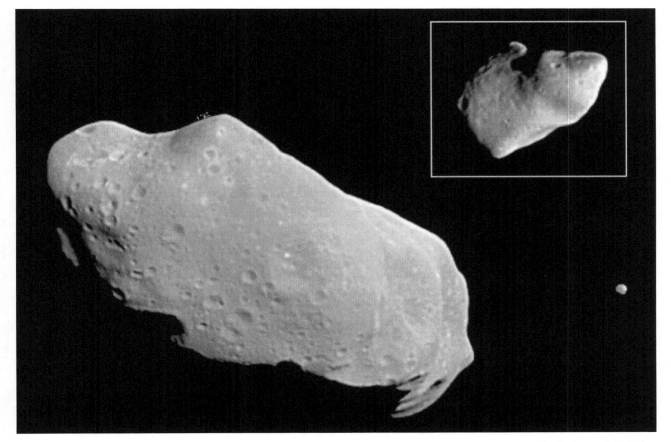

Asteroid Ida, so named for its potato shape, as imaged by the Galileo spacecraft. Ida's satellite Dactyl appears in the inset., (NASA/JPL/USGS)

nuclear device as the only viable way to avert an asteroid impact. In many real cases, nuclear explosions detonated within, on the surface of, or close to asteroids either would be insufficient or could merely fragment it so badly that an even worse situation, a swarm of impacting bodies, might ensue.

Methods of Study

Studies of asteroids hold the potential for expanding our understanding of the formation of bodies of sizes between the smallest objects and full planets, and also could lead to development of technology to prevent an impact that might devastate life on Earth and even wipe out civilization.

The Galileo spacecraft passed near enough to two asteroids to photograph them directly. The NEAR spacecraft orbited Eros and later landed on its surface, providing close-up photographs of an asteroidal surface. For the most part, however, indirect methods of remote sensing must be used to determine asteroidal properties by studying the reflected electromagnetic radiation that comes from their surfaces. These methods include photometry, infrared radiometry, colorimetry, spectroscopy, polarimetry, and radar detection. They can be augmented by comparative studies with meteorites, whose composition and structure can be analyzed by direct methods in the laboratory. Such methods include chemical, spectroscopic, and microscopic analysis, and processes of fragmentation can be studied by producing high-speed collisions between comparable materials in the laboratory. Such comparative studies must recognize various differences between meteorites and asteroids. The masses of only the three largest asteroids have been determined from their gravitational effects on other bodies; their densities are between 2.3 and 3.3 grams per cubic centimeter.

Photometry is the study of how light is scattered by various surfaces. The varying brightness of reflected sunlight from asteroids can be measured by photoelectric observations to determine their rotation periods and approximate shapes. One test of this method was made in 1931, when the Amor asteroid 433 Eros came close enough (23 million kilometers) for scientists to observe the tumbling motion of this elongated object (7 by 19 by 30 kilometers) and to confirm its 5.3-hour rotation. The size of an asteroid can be estimated from its brightness together with its distance, orbital position, and albedo. The albedo is important, since a bright, small object may reflect as much light as a dark, large object. Since a dark object absorbs more heat than a light object, albedos can be determined

by comparing reflected light with thermal radiation measured by infrared radiometry. Photometric measurements also give information on surface textures. Colorimetry involves measuring the range of wavelengths in the reflected light to determine surface colors. Most asteroids are either fairly bright, reddish objects (with albedos of up to 23 percent) composed largely of silicate-type materials or grayish objects, at least as dark as the Moon (11 percent albedo), composed of carbonaceous materials.

Spectroscopy is the spectral analysis of light and can be used to infer the composition of many asteroids. Optical and infrared reflectance spectra exhibit absorption bands at characteristic frequencies for given materials. An asteroid's surface composition is determined by comparing its spectrum with the spectra of light reflected from meteorites of known composition. Examples of this method applied to U-type (unclassified) asteroids include the identification of the silicate mineral pyroxene in the infrared spectrum of Apollo asteroid 1685 Toro, and the matching of the surface of Vesta with a basaltic achondrite that resembles lava. Most asteroids appear to have unmelted surfaces with little or no evidence of lava eruptions. About two-thirds of the Trojans are D-type asteroids with no known meteorite counterparts because of their distance from Earth. Their spectra have been matched with the spectra of coal-tar residues, suggesting possible organic compounds.

Polarimetry uses measurements of the alignment of electric field vibrations of the reflected sunlight and its variation with direction to estimate albedos. Polarization measurements have also been interpreted as evidence for dust-covered surfaces, but they leave uncertainty about the depth of the dust layer. Radar observations of Eros during a close approach to Earth in 1975 were made at a wavelength of 3.8 centimeters and indicated that the surface must be rough on a scale of centimeters. Since optical polarimetry suggests that Eros is dusty, the radar results imply that the dust must be too thin to smooth rock outcrops of more than a few centimeters. Radar measurements also provided independent estimates of the size of Eros, confirming photometric estimates of its dimensions. The NEAR spacecraft confirmed these observations.

As spacecraft results such as this demonstrate, the best method to study asteroids is by means of a space probe. When Pioneers 10 and 11 passed through the asteroid belt, scientists found that it has no more dust than any other part of the solar system. The Galileo probe encountered Gaspra in 1991 and Ida in 1993, both S-type asteroids. The probe determined the masses, sizes, and shapes. The

The Torino Asteroid Impact Hazard Scale

Scale	Description
0	Event having no likely consequences: The likelihood of a collision is zero, or low enough to be effectively zero. This designation also applies to any small object that, in the event of a collision, is unlikely to reach the Earth's surface intact.
1	Event meriting careful monitoring: The chance of collision is extremely unlikely, about the same as a random object of the same size striking the Earth within the next few decades.
2	Event meriting concern: A somewhat close but not unusual encounter. Collision is very unlikely.
3	Event meriting concern: A close encounter, with 1 percent or greater chance of a collision capable of causing localized destruction.
4	Event meriting concern: A close encounter, with 1 percent or greater chance of a collision capable of causing regional devastation.
5	Threatening event: A close encounter, with a significant threat of a collision capable of causing regional devastation.
6	Threatening event: A close encounter, with a significant threat of a collision capable of causing a global catastrophe.
7	Threatening event: A close encounter, with an extremely significant threat of a collision capable of causing a global catastrophe.
8	Certain collision: A collision capable of causing localized destruction. Such an event occurs somewhere on Earth between once per 50 years and once per 1,000 years.
9	Certain collision: A collision capable of causing regional devastation. Such an event occurs between once per 1,000 years and once per 100,000 years.
10	Certain collision: A collision capable of causing a global climatic catastrophe. Such an event occurs once per 100,000 years or less often.

Cassini spacecraft on its way toward orbit about Saturn flew through the asteroid belt and passed asteroid 2685 Masursky at a distance of 1.6 million kilometers. Named after the famed planetary scientist Hal Masursky, this body was a little understood 15- by 20-kilometer asteroid prior to Cassini's encounter.

Before the Pioneer 10 and 11 passages there were serious concerns that spacecraft might not be able to pass safely through the main asteroid belt. Much has been learned about the density of material in the belt since the space age began. Thus far, no spacecraft sent into the belt has experienced serious damage from an impact with asteroidal material or an actual asteroid body. Minor hits on dust detectors have been recorded, however. Robotic spacecraft investigations have provided much information about the nature of the various types of asteroids, as has analysis of meteorites found on Earth that are believed to have come from certain asteroids.

The NEAR spacecraft was launched on February 17, 1996, and was directed toward a rendezvous with the asteroid Eros three years later. Eighteen months out from Earth, NEAR flew by the asteroid Mathilde. It successfully reached Eros, and for well over a year NEAR orbited Eros at varying altitudes providing high-resolution images of the surface of this S-type asteroid. After completing its primary mission, NEAR gently touched down on Eros on February 12, 2001. A total of 69 high-resolution images of the asteroid's surface were taken on the way down during a soft-contact landing. The final picture was taken at an altitude of 130 meters and covered an area of 6 meters by 6 meters. Within that final frame was a portion of a 4-meter-wide boulder, as well as evidence of a dusty surface pocked with small rocks and tiny craters. Much to the surprise of the Johns Hopkins University Applied Physics Laboratory research team controlling the spacecraft, NEAR survived its landing and transmitted data back to Earth for two weeks before falling silent. The team was lucky in that the spacecraft's antenna pointed toward Earth and the solar arrays faced partially toward the Sun after impact.

The next step in spacecraft-based investigations of asteroids is the Dawn mission, a robotic probe designed to orbit two different bodies. Dawn's mission is to visit the two largest asteroids, Ceres and Vesta. By comparison with Ceres, the asteroid upon which NEAR settled was a tiny speck. Ceres is a spherical body with a diameter of 960 kilometers. Indeed under an official review of classification for solar-system objects, Ceres is now officially designated a dwarf planet—a characterization it shares (much to the displeasure of many in the scientific community) with Pluto, which was demoted from full planet status to that of a dwarf. To accomplish its mission on a minimum of propellant, Dawn is outfitted with ion propulsion similar to that demonstrated by the Deep Space 1 spacecraft. To achieve its science goals, Dawn is outfitted with a framing camera, a mapping spectrometer, and a gamma-ray and neutron spectrometer. The goal is to image the surface of these two large asteroids and to determine their composition.

Dawn launched on September 27, 2007, and was set up for a gravity assist from Mars in early 2009. Arrival at Vesta was planned for September, 2011. The ion propulsion system would then break Dawn out of Vesta orbit in April, 2012, and send the spacecraft toward a rendezvous with Ceres in February, 2015. Assuming the spacecraft remains healthy and propellant is available when the primary mission ends in July, 2015, Dawn could be redirected to other asteroids within reach.

Samples of rocks believed to come from various portions of the asteroid belt fall on Earth regularly and have been subjected to intense study. The next step in asteroid investigation would be to return pristine samples of asteroids so the asteroid samples are not altered on their outer layers by passage through Earth's atmosphere. A robotic mission to collect and then return samples from any asteroid is possible with contemporary technology.

Perhaps the greatest potential for insight into the nature of asteroids would be a human expedition to such a body. Shortly after the adoption of the Vision for Space Exploration in 2004, National Aeronautics and Space Administration (NASA) entertained the possibility of sending a crewed Orion Crew Exploration vehicle into deep space for a rendezvous with an asteroid. In terms of propulsion requirements, it is slightly less intensive to send a piloted spacecraft to a near-Earth asteroid than to the Moon. Such a mission could take at least six months to reach a target and up to another year to return to Earth. It could return to Earth large amounts of carefully selected asteroid samples for detailed analysis. The potential for gathering information that might be used someday to divert an asteroid that threatened to impact Earth would be tremendous.

Context

Asteroids usually cannot be seen with the unaided eye, but they provide important clues for understanding planet formation: They can have major effects on the Earth and, in fact, have had such effects during the planet's history. At one time, it was assumed that the asteroid belt was formed by the breakup of a planet between Mars and Jupiter. However, the combined mass in the belt is much less than that of any planet (only 0.04 percent Earth's mass), and the observed differences in the composition of asteroids at different locations in the belt make it unlikely that they all came from the same planet-sized object. It now appears that asteroids are original debris that was left over after planet formation and that has undergone complex processes such as collisions, fragmentation, and heating. Apparently, strong tidal forces caused by Jupiter's large mass prevented small bodies between it and Mars from combining to form a single planet in their region.

It appears, therefore, that asteroids are among the oldest objects in the solar system, left over from the time immediately before planet formation concluded. Studies of these objects should provide clues to the structure and composition of the primitive solar nebulae. Different types of asteroids found in different regions of the solar system support the theory of planetesimal origin through a sequence of condensation from a nebular disk around the Sun. Asteroids farther from the Sun, beyond the main belt, may have contained more ice; those that formed closer, within the belt, may have been primarily stony or stony-iron materials. Some of these planetesimal precursors of asteroids were probably perturbed during close passes by neighboring planets into elongated Apollo-like orbits that cross Earth's orbit. Other objects on similar orbits may have been comets that remained in the inner solar system long enough to lose their volatile ices by evaporation. Processes of collision and fragmentation among these objects provide direct evidence about the earliest forms of matter.

Special interest in Apollo asteroids arises from their potential for Earth collisions. Objects as small as 100 meters hit Earth about once every two thousand years, and the 30 percent that fall on land can produce craters a kilometer in diameter. Such impacts would devastate much wider areas by their shock waves, and dust thrown into the upper atmosphere could have marked effects on climate. Growing evidence suggests that asteroid collisions in the past might have contributed to major extinctions of species, such as the dinosaurs, and perhaps even caused reversals of Earth's magnetism. Thin layers of iridium, often found in meteorites, have been identified in Earth's crust at layers corresponding to such extinctions. Satellite photography has revealed about one hundred apparent impact craters on Earth with diameters up to 140 kilometers. It is likely that many more succumbed to processes of erosion. Knowledge of Apollo orbits might make it possible to avoid such collisions in the future.

Asteroids also offer the possibility of recovering resources with great economic potential. Some contain great quantities of nickel-iron alloys and other scarce elements; others may yield water, hydrogen, and other materials useful for space-based construction. Estimates of the economic value of a kilometer-sized asteroid reach as high as several trillion dollars. A well-designed approach to space mining might someday help to take pressure off Earth's ecosystem by providing an alternative to dwindling resources, and space-borne manufacturing centers might alleviate pollution on Earth.

Joseph L. Spradley and David G. Fisher

Further Reading

Barnes-Svarney, Patricia. *Asteroid: Earth Destroyer or New Frontier?* New York: Basic Books, 2003. In-depth coverage of technical issues about asteroids necessary to understand the danger that an impact on Earth represents. Makes connections to science-fiction stories involving such disasters and allows the reader the ability to determine what is often incorrectly portrayed in doomsday documentaries and fiction with asteroid impact themes.

Beatty, J. Kelly, Carolyn Collins Petersen, and Andrew Chaikin, eds. *The New Solar System.* 4th ed. Cambridge, Mass.: Sky, 1999. A richly illustrated summary of early space-age discoveries that radically revised knowledge of the solar system. Discusses the various types of asteroids.

Bobrowsky, Peter T., and Hans Rickman, eds. *Comet/Asteroid Impacts and Human Society: An Interdisplinary Approach.* New York: Springer, 2007. Suitable for an interdisciplinary college course at the introductory level about the science and societal issues related to an impact on Earth by a near-Earth asteroid or comet.

Gehrels, Tom, ed. *Asteroids.* Tucson: University of Arizona Press, 1979. A classic, authoritative and comprehensive book on asteroids available in English. It contains about fifty articles on every aspect of asteroid research, including extensive references to original research papers. Most articles are technical, but the first seventy-five pages provide a readable introductory survey. Tabulations in the last section provide data of various kinds on all asteroids that have been studied.

Harland, David H. *Jupiter Odyssey: The Story of NASA's Galileo Mission.* New York: Springer Praxis, 2000. Provides virtually all of NASA's press releases and science updates during the first five years of the Galileo mission in a single work, including Galileo's encounters with asteroids. Includes an enormous number of diagrams, tables, lists, and photographs.

Hartmann, William K. *Moons and Planets.* 5th ed. Belmont, Calif.: Thomson Brooks/Cole, 2005. An updated version of a classic text that covers all aspects of planetary science. Results for the entire NEAR mission are presented. Additional material relating to asteroids is included in chapters on comets, meteorites, planetary evolution, and cratering. An appendix on planetary data includes some asteroid data for comparison, and an extensive bibliography includes about seventy entries on asteroids.

Lewis, John S. *Rain of Iron and Ice: The Very Real Threat of Comet and Asteroid Bombardment.* New York: Basic Books, 1997. A comprehensive survey of meteorites, impact-cratering processes, and the concept of cataclysm. About the latter, the book provides a historical look at the change in scientists' belief in uniformitarianism to their recognition of catastrophism. Plans for preventing a major impact are discussed.

Time-Life Books. *Comets, Asteroids, and Meteorites.* Alexandria, Va.: Author, 1990. Heavily illustrated in color, offers an excellent collection of photographs of comets, pictures of meteorites, and descriptions of asteroids.

ASTEROID TOUTATIS PASSES NEAR EARTH

Category: Small Bodies

Discovered and named in 1989, the asteroid Toutatis, which passes near Earth roughly every four years, not only promises astronomers unprecedented access to data concerning asteroid formation but also has generated frank discussions concerning possible scenarios should an object of that magnitude ever collide with Earth.

Overview

On January 4, 1989, asteroid 4179 was discovered by French astronomer Christian Pollas while he was working at the Observatoire de la Côte d'Azur at Caussols, in southern France. Pollas, a veteran astronomer credited with discovering numerous asteroids, happened to spot the bright, fast-moving object on photographic plates that had been taken in an effort to measure the telemetry of Jupiter's obscure satellites. Pollas named the asteroid Toutatis after a powerful, protective god of fertility, war, and prosperity common to both Gallic and Celtic mythologies. As Pollas found out, however, Toutatis was also a deity figure in Les Aventures d'Asterix, a hugely successful, long-running French comic book series set in medieval Europe; ironically, in the series, the tribe that worships Toutatis is convinced the sky is soon to fall. Once the new asteroid was identified, its approach was tracked by the scientific community, which confirmed its Earth-crossing orbit, its speed (roughly 67,112 miles per hour), and its dimensions (2.9 miles long, 1.5 miles wide, 1.2 miles thick, considerable for asteroids).

Apart from the usual scientific buzz that inevitably accompanies the identification of any new heavenly body, particularly a near-Earth object, the discovery of Toutatis brought additional excitement because, given the asteroid's location—just inside the Earth's orbit out to the main asteroid belt between Mars and Jupiter—it would return to the Earth's observation roughly every four years. Further, its particularly low orbit would bring the asteroid close to the Earth—relative to space measurement. In 1992, for instance, the asteroid came within 2.5 million miles; in November, 1996, 3.3 million miles; and in September, 2004, just over 1 million miles, the closest approach of any known object until 2060. Such regular "close" brushes have enabled astronomers unprecedented access to the asteroid using only Doppler ground radar without having to launch satellites to obtain the data. Following the 1996 pass, for instance, the NASA Jet Propulsion Laboratory in Pasadena, California, released startlingly vivid delay-Doppler radar images of the asteroid's pocked surface. Indeed, in its 2004 flyby, Toutatis was visible for backyard sky watchers with only binoculars.

Thus, within the relatively brief time Toutatis has been known, scientists have been able to study it. They have found that it is quite an unusual object, which led, in turn, to alarmist concerns. The images of Toutatis revealed that it did not possess an asteroid's usual spherical shape; rather, it is oblong with a peanut-shell shape, with one lobe substantially smaller than the other. Astronomers conjecture that, given the asteroid's heavily cratered face, Toutatis was once two asteroids that slammed together in a violent fusion.

Further intriguing astronomers was Toutatis's irregular rotation: Asteroids spin in a tight and predictable spiral, like a child's top, along a single axis of rotation; but Toutatis spins in a wobbly, tumbling motion—the result, astronomers theorize, of numerous collisions with other floating debris. Consequently, Toutatis does not have a fixed pole of rotation; rather, it follows what is termed non-principal axis rotation—quite a rare occurrence. Toutatis maintains two entirely different rotation motions, which means that if a person could actually stand on the asteroid, he or she would see the Sun rise and set along a different path each day. Thus the asteroid does not maintain a fixed "day"; rather, it completes its spin sometimes in 5.4 days and sometimes in 7.3 days, a period that in either case is far slower than those of most asteroids.

The implications for scientific investigation were promising: Given the asteroid's irregular rotation, its quadrennial flybys, and its slow tumbling movement, Toutatis would expose virtually its entire circumference at one time or another for Earth observation. The scientific community came to view the regular approach of Toutatis as an opportunity to advance theoretical explanations for the beginnings of the universe, as asteroids are widely believed to be debris left over from initial cosmic eruptions that first forged the planets and the stars. Asteroids' mineral contents are particularly helpful in directing such theories. Given Toutatis's relatively close orbit, the scientific community has raised the possibility of launching a robotic exploratory vehicle to engage the asteroid and to determine its exact mineral makeup during one of three approaches in the 2020's.

The very characteristics of Toutatis's orbit that have so intrigued scientists, however, have caused alarm in the lay

A computer-generated view of Earth as seen from Asteroid Toutatis. (NASA)

community. Such alarm was perhaps inevitable when, because of its orbit, Toutatis was pro forma designated early on as a "potentially hazardous asteroid" (PHA). Asteroids have been known to collide with the Earth's surface. Most recently, a devastating strike took place in 1908 in Tunguska, Russia, in which an asteroid much smaller than Toutatis leveled more than 700 square miles of Siberian wilderness. An asteroid roughly twice the size of Toutatis is believed to have caused the extinction of the dinosaurs more than sixty million years ago. Astronomers have conjectured that asteroids regularly collided with the Earth during its earliest eons of formation and that the cratered face of the Moon verifies the impacts of such collisions. Given the traffic of asteroids currently in the solar system (conservative estimates suggest more than 300,000 such objects of at least 300 feet long), scientists have long held that collisions with the Earth are inevitable, although they predict such collisions would occur only every thousand years or so.

The topsy-turvy orbital patterning of Toutatis, however, means that scientists cannot confidently predict the exact path of Toutatis's approaches beyond several centuries. In fact, given the asteroid's unusually low inclination—less than half a degree from Earth's—in several computer scenarios played out six centuries into the future, Toutatis is tracked to actually collide with the Earth, encouraging lay speculation to dub the mountainous asteroid the "Doomsday Rock." Such speculation routinely describes Toutatis's eccentric orbit as "unpredictable," although scientists quickly point out the significant difference between an "unpredictable" orbit and Toutatis's "irregular" orbit, which is unique but definitely patterned.

Nevertheless, within the nonscientific community (fed by Internet hype), each approach of Toutatis triggers a considerable volume of alarmist misinformation suggesting that the asteroid's close pass confirms anxieties about an apocalyptic collision with the Earth, although astronomers are quick to point out that "close" still assures millions of miles between the Earth and the asteroid. The asteroid's 1996 flyby, however, inspired two disaster movies: the 1998 blockbuster *Armageddon*, in which a emergency space mission must detonate an approaching asteroid, and, that same year, the less successful *Deep Impact*, in which a chunk of a rogue comet actually strikes the Earth and triggers monumental destruction on the East Coast of the United States.

Significance

Although Toutatis offers a rare opportunity for astronomical observation, it has also occasioned a concerted effort by an international cartel of scientists, astronauts, diplomats, insurance executives, lawyers, and astrophysicists to petition the United Nations to draft a specific protocol for addressing the threat of a collision. Despite the mathematical certainty that an asteroid of considerable dimension will collide with the Earth, no framework exists setting out a concrete course of response.

In 2007, a U.N. blue-ribbon committee submitted a draft of just such a global policy, including recommendations for determining which governments would be charged with directing attempts to deflect any incoming objects and a proposed structure for international relief operations should a catastrophic collision occur. Debate on the protocol is set to conclude by 2009. Given the enormous reach into the visible universe afforded by computer-enhanced observation technologies and the detection over the last decade of many new PHAs—as many as 20,000 of which have yet to be exactly identified—the National Aeronautics and Space Administration was charged by Congress in 2005 to increase its efforts agressively both to identify asteroids that could pose a threat to the Earth (only one hundred such objects had been cataloged by 2007) and to investigate feasible strategies of preparation should the planet face such a threat.

Joseph Dewey

Further Reading

Bobrowsky, Peter T., and Hans Rickman, eds. *Comet/Asteroid Impacts and Human Society: An Interdisciplinary Approach.* New York: Springer, 2007. Measured analyses that project the immense cultural and social alterations in the event of an impact.

Hallam, Tony. *Catastrophes and Lesser Catastrophes: The Causes of Mass Extinction.* New York: Oxford University Press, 2005. From a distinguished geologist, a look at historic evidence of catastrophes including asteroid impacts. Closes with a review of current international efforts—scientific, technological, and political—to confront such collisions.

Palmer, Trevor. *Perilous Earth: Catastrophe and Catastrophism Through the Ages.* New York: Cambridge University Press, 2003. Written for a general audience but grounded in meticulous research, the book explores the differences between preparation and panic.

Spangenburg, Ray, and Kit Moser. *If an Asteroid Hit the Earth.* New York: Franklin Watts, 2000. Sobering analysis of the effects of an asteroid hit. Includes helpful context concerning past collisions.

Verschuur, Gerrit L. *Impact! The Threat of Comets and* Asteroids. New York: Oxford University Press, 1997. Landmark investigation into historic collisions and their impact, particularly the Tunguska catastrophe.

CALENDARS

Category: Space, Time, and Distance.

Astronomical observations led to the development of various calendars that use different methods of resolving the need for "leap" days, months, or years.

Overview

Even the earliest human beings must have noticed the astronomical cycles: the alternation of day and night, the pattern of the changes in the moon's shape and position, and the cycle of the seasons through the solar year. It must have been frightening every autumn as the days became shorter, causing concern that the night might become permanent. This led to celebrations of light in many areas as the days began to lengthen again. Once the repetitions of the patterns were recognized, people could count them to keep track of time. Longer cycles helped avoid difficulties in keeping track of large numbers—once approximately 30 days had been counted, people could, instead, start counting "moons." This same technique of grouping also occurred in the development of counting systems in general—leading to place-value structures in numeration systems.

The problem was that the shorter cycles did not fit evenly into the longer cycles. Trying to fit the awkward-length cycles together actually led to some mathematical developments: two different cycles would come together at the least common multiple of the lengths of their cycles; modular arithmetic and linear congruences were methods of handling leftover periods beyond the regular cycle periods.

A 1412–1416 illumination depicting the month of March with the constellations of the zodiac on top. (Photos.com)

The Julian and Gregorian Calendars

The Romans developed the Julian calendar (named for Julius Caesar), recognizing that the exact number of 365 days in one year was slightly too short and would soon throw the calendar off the actual cycle of the solar year. They found a remedy by assuming the solar year to be 365.25 days. To handle the one-quarter day, they added one full day every four years—the day that we call "leap-year day" on February 29 of years whose number is a multiple of four. This gives, days in four years, or an average of 365.25 days per year as desired. However, the actual solar year is 365.2422 days long (to four decimal places),

about 11 minutes less than the Romans' value. Even in a human lifetime, this is negligible. Over centuries, however, the extra time builds up so that by the 1500s, the calendar was 10 days off from the solar cycle (for example, the vernal equinox seemed to be coming too late).

In 1582, Pope Gregory XIII assembled a group of scholars who devised a new system to fit better. It kept the Roman pattern except that century years (1600, 1700), which should have been leap years in the Roman calendar, would not have a February 29 unless they were multiples of 400. For example, 1900 was not a leap, year but 2000 was. In the full 400-year cycle, there are (400×365) regular days + 97 leap-year days = 146,097 days, making an average of 365.2425 days per year. This cycle is only .0003 days (about 26 seconds) too much; in 10,000 years, we would gain three extra days. This system was called the Gregorian calendar. Since the longer Julian calendar had fallen behind the solar year by about 10 days, the changeover to the Gregorian required jumping 10 days.

Various countries in Europe changed at different times. with each switch causing local controversy as people felt they were being "cheated" out of the skipped days. The effects of the change are noticed in history. When Isaac Newton was born, the calendar said it was December 25, 1642; but later England changed the calendar, so some historians today give Newton's birthday as January 4, 1643. The Russians did not change their calendar until after the 1917 October Revolution, which happened in November by the Gregorian calendar.

The Lunar Calendar

The other incongruity of calendar systems is that the moon cycle of 29.53 days does not fit neatly in the 365.2422 days of the year. Twelve moon periods is 11 days shorter than a year, and 13 "moons" is 18 days too long. It is interesting to note that of the three major religious groups of the Middle East—the Christians, the Muslims, and the Jews—each chose a different way to handle "moons/months." The Christians (actually, originally, the Romans) ignored the moon cycle and simply created months of 30 and 31 (and 28 or 29) days. The Muslims considered their year to be 12 moon cycles and ignored the solar year. This means that dates in the Muslim calendar are shifted back approximately 11 days each year from the solar calendar, and Muslim festivals move backward through the seasons.

People in the Jewish faith chose to keep both the solar and lunar cycles. After 12 lunar months, a new year begins—as in the Muslim calendar—11 days "too early."

$$3\left(365\right)+366=1461$$

However, after the calendar slips for two or three years—falling behind the solar calendar by 22 or 33 days—an extra month is inserted to compensate for the loss. There is a 19-year pattern of the insertion of extra months, which keeps the year aligned with the solar year. Interestingly, the traditional east Asian calendar follows a pattern very similar to the Jewish calendar.

The Mayan Calendar

The Mayans of Central America had a very complex pattern of cycles leading to a 260-day year for religious purposes, and a regular solar year that was used for farming and other climate-related activities. Their base-20 numeration system, which should have had place-value columns of 1-20-400-8000, was adjusted to 1-20-360 to fit into the 365+ days of the year. They were also notable for developing massive cycles of years lasting several millennia, including one ending in late 2012 of the Gregorian calendar.

Lawrence H. Shirley

Further Reading

Aslaksen, Helmer. "The Mathematics of the Chinese Calendar." http://www.math.nus.edu.sg/aslaksen/calendar/chinese.shtml.

Crescent Moon Visibility and the Islamic Calendar. http://aa.usno.navy.mil/faq/docs/islamic.php.

Duncan, David Ewing. *Calendar: Humanity's Epic Struggle to Determine a True and Accurate Year*. New York: Harper Perennial, 2001.

Rich, Tracey R. "Judaism 101: Jewish Calendar." http://www.jewfaq.org/calendar.htm.

Richards, E. G. *Mapping Time: The Calendar and Its History*. New York: Oxford University Press, 2000.

Stray, Geoff. *The Mayan and Other Ancient Calendars*. New York: Walker & Company, 2007.

CERES

Category: Small Bodies

Discovered in 1801, Ceres, named for the Roman goddess of agriculture, is the largest of the main-belt asteroids. For a short time it was believed to be the eighth planet in the solar system, but discovery of additional large main-belt asteroids influenced astronomers to revoke its planetary status. Discoveries of Pluto-sized objects in the Kuiper Belt have once again brought Ceres back into the discussion of what constitutes the definition of a planet.

Overview

On the night of January 1, 1801, the Italian astronomer Giuseppe Piazzi was observing the heavens when he noted a faint object that did not appear on his star charts. At first he thought it might be a comet, but it did not have the typical "fuzzy" appearance associated with comets. If not a comet, what could it be? By observing its motion over the next several weeks, Piazzi was able to determine that its orbital speed was greater than that of Mars but slower than that of Jupiter. This suggested to him that the object must lie between the orbits of Mars and Jupiter.

Additional help came from the German mathematician Carl Friedrich Gauss, who had perfected a means of calculating orbital motion based on limited observations. When he applied his method to Piazzi's observations, Gauss was able to calculate where and when this mysterious object should next appear, and it did just as he predicted. Within one year of Piazzi's discovery, Heinrich Olbers and Franz von Zach were able to relocate Ceres and refine its 4.6 Earth-year orbit. Later scientists were able to determine that it has a spherical shape with a 930-kilometer diameter. In comparison to the Moon, Ceres is one-third its size, but with only less than 2 percent of its mass, giving it a much lower density of 2.1 grams per cubic centimeter.

Piazzi's observations and the calculations of Gauss led many of the leading scientists of that time to believe that a new planet had been discovered. This conclusion seemed logical, based on an earlier idea first suggested by Johann Daniel Titius of Wittenberg and later championed by Johann Elert Bode. In 1792, Bode pointed out an apparent mathematical relationship between the distances of the various planets to the Sun. He suggested that the planets were positioned at specific distances from each other based on a mathematical ratio that would later be referred to as Bode's law. This worked reasonably well for all the planets from Mercury through Uranus, with the

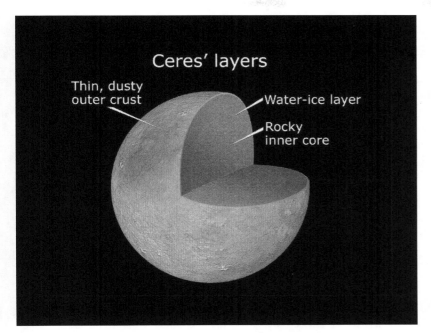

Ceres' layers

Thin, dusty outer crust

Water-ice layer

Rocky inner core

An artist's conception of the interior of Ceres, formerly classified as an asteroid and now the only dwarf planet in the inner solar system. (NASA/ESA/A. Field, STScI)

exception of an apparent gap between Mars and Jupiter. When Piazzi found his mystery object positioned in this gap where Bode suggested a planet should be, this seemed to be the observational confirmation of Bode's law. Even though modern science treats Bode's law as more of an interesting coincidence rather than a scientific law, it did serve a purpose at that time and contributed to the eventual discovery of Neptune.

Although initially proclaimed the eighth planet in 1801, Ceres did not long retain its planetary status. The excitement created in the astronomical community by the discovery of Ceres led to a systematic search of the heavens, which centered on the plane of the ecliptic. Scientists believed that many new and interesting objects would soon be found, and they were right. Within the next six years, three new asteroids—Pellas, Juno, and Vesta—were found within the same general vicinity as Ceres. With four minor bodies now occupying the same region of space, scientists concluded that no one planet would fill the gap in Bode's law. A new theory would have to be created to explain the presence of so many small bodies occupying planetary positions.

Since that time, many theories have been created to explain the presence of the main-belt asteroids. One of the more popular, but incorrect, theories described a

large planet exploding and creating a huge number of smaller bodies ranging from the largest, Ceres, down to extremely small meteoroid-sized fragments. Perhaps the most widely accepted theory now describes a "planet that never formed." This theory can be supported by the generally accepted nebular hypothesis of planetary formation, which envisions a final stage of accretion during which a huge number of smaller bodies are attracted to each other and form into a much larger object. In the case of Ceres and the other main-belt asteroids, that final stage was interrupted and they never fully accreted into a single large object.

Beginning in the late twentieth century, the study of Ceres and its family of asteroids was no longer regulated by the limitations of Earth-based telescopic observations. The Hubble Space Telescope operating well above the Earth's atmosphere revealed details never before seen by surface-based telescopes. In addition, modern astronomers have the opportunity to send their spacecraft-borne instruments directly to the asteroids to get close-up views of their surfaces and analyze their mineralogical compositions. Several flyby spacecraft missions have investigated a number of smaller asteroids revealing previously unimagined surface conditions. One in particular, the NEAR Shoemaker probe, first orbited and then actually landed on Eros, giving scientists their first detailed images from the surface of an asteroid. The Japanese probe Hayabusa is believed to have landed on the surface of the asteroid Itokawa and collected a sample for return to Earth. Data returned by these missions have rewritten the textbooks on what is known about asteroids. The Dawn spacecraft, en route to both Ceres and Vesta, was designed first to orbit Vesta in 2011 and then leave orbit and go on to rendezvous with and orbit Ceres in 2015. Dawn's mission was to collect sufficient data to help scientists gain a better understanding of the conditions present at the initial accretion stage of planetary formation in the solar system.

Ceres is very different from Vesta. Studies based on a comparison of densities and surface reflectivity have shown that Ceres may have what is considered to be a "wet" surface, composed of water-bearing minerals, as opposed to the "dry" surface minerals of Vesta. Some scientists speculate that Ceres may have a total amount of water locked up in its interior that could rival that of the Earth's surface, while Vesta is more comparable to Earth's moon. Other Dawn experiments should shed light on the internal structure, shape, composition, and mass of these two primordial bodies. Scientists have been able to determine that a particular class of meteorites, the HED achondrites, is probably derived from Vesta. It is believed that most meteorites are fragments of crustal material that was blasted off an asteroid's surface during the accretion process or from later impacts. This is based on reflectivity studies of minerals present on the surface of Vesta and from radioisotope chronology studies of the HED meteorites. All evidence seems to point to Vesta. No such data exists for Ceres. This is why a spacecraft mission to Ceres is so important.

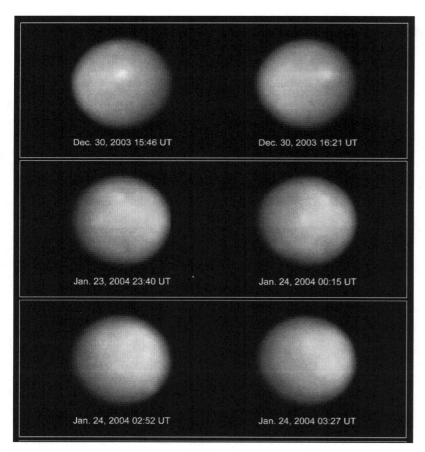

Released in 2005, these six images of Ceres were captured by the Hubble Space Telescope between December, 2003, and January, 2004. (NASA/J. Parker et al., Southwest Research Institute)

Knowledge Gained

The discovery and later scientific study of the asteroid Ceres had a profound effect both on our early view of the solar system and on our subsequent understanding of the origin and nature of planetary bodies. At the time of its discovery (1801), Ceres represented another "new" object in the heavens that was unknown to the ancient astronomers. It had been less than fifty years since the return of Halley's comet (1758) had literally galvanized the concept of gravity and its effects on motion, and only twenty years after the discovery of Uranus (1791). The discovery of three other asteroids would soon follow, but after 1807, no other asteroid was detected until 1845.

The following year Neptune was discovered, and astronomers had definitely moved into the "modern era" with bigger and better telescopes, technology that included spectroscopy and photography, and a more scientific perspective of the universe. With Neptune recognized as the eighth planet, the asteroids fell into their proper place within the structure of the solar system. The study of asteroids remained within the domain of observational astronomers until scientists and engineers could develop the technology to send their scientific instruments to the planets and minor bodies. Once this happened, the outpouring of data changed our view of the nature and origin of the planets.

To the early astronomers Ceres was only a faint speck of light in the night sky. Today modern astronomers see it quite differently, primarily as a result of the observations made by the Hubble Space Telescope's Advanced Camera for Surveys. In 2005 astronomers observed Ceres through a complete nine-hour revolution, taking 267 photographic images. From these observations astronomers were able to determine that Ceres has a spherical shape with a diameter slightly wider at the equator than at the poles. This suggests that it has a differentiated internal structure with denser materials forming a core and lighter materials closer to the surface. Scientists suspect that, because Ceres' density is much lower than Earth's, large amounts of water ice may exist either on the surface of Ceres or buried within its crust. This surmise is supported by spectral evidence for water-bearing minerals that may be present on the surface that are not representative of Ceres' crystal rocks. Additional microwave studies suggest that this surface material might be dry clay. All considered, Ceres could turn out to be the most Earth-like body in the solar system; it may even be a haven for primitive forms of life.

Context

Clearly the asteroids hold many vital clues to unraveling the mysteries surrounding the formation of the planets. On the basis of their respective sizes, densities, and chemical compositions, the planets in our solar system are divided into three major groups: the terrestrial (Earth-like) planets, the Jovian (Jupiter-like) planets, and the dwarf planets. The third group can be further divided into rocky objects like Ceres and into ice bodies like Pluto. These dwarf planets, rocky or ice, most likely represent a fundamental primordial stage in the formation of planets.

By studying these early remnants of planetary formation, scientists can achieve a clearer picture of their formative processes. Each group will have its own distinctive secrets to reveal. The rocky dwarf planets positioned between Mars and Jupiter formed under conditions of higher temperature, higher density, and higher velocity than the icy worlds at the edge of the solar system in the Kuiper Belt. It is believed that at this distance from the Sun, these objects have remained essentially unchanged over the last 4.6 billion years. In 2015, the New Horizons spacecraft will visit Pluto and send back images and data giving science its first close-up look at this unknown world. In that same year the Dawn spacecraft will orbit Ceres. Perhaps then, with both worlds under study, science will be able to fill in many of the gaps in our understanding of planetary formation.

Paul P. Sipiera

Further Reading

Bell, Jim, and Jacqueline Mitton, eds. *Asteroid Rendezvous: NEAR Shoemaker's Adventures at Eros*. Cambridge, England: Cambridge University Press, 2002. A collection of nine scientific articles that provide the reader with an overview of an asteroid rendezvous mission and what to expect from the anticipated Dawn mission to Ceres and Vesta. Suitable for a wide range of readers.

Bottke, William F., Jr., Albertoi Cellino, Paolo Paolicchi, and Richard P. Binzel, eds. *Asteroids III*. Tucson: University of Arizona Press, 2002. This comprehensive work is a compilation of scientific papers that cover virtually every aspect of asteroid research. The paper on Piazzi and Ceres is especially relevant. Best suited for the graduate student and professional scientist.

Hartmann, William K. *Moons and Planets*. 5th ed. Belmont, Calif.: Thomson Brooks/Cole, 2005. An excellent reference for a variety of topics in planetary science, including asteroids. Suitable for readers of high school-level and above.

Kowal, Charles T. *Asteroids: Their Nature and Utilization*. Chichester, England: Ellis Harwood, 1988. A good basic reference source for general information about asteroids, including both their potential for commercial use and their potential as a hazard to Earth. Suitable for both astronomy enthusiasts and students at the undergraduate and graduate levels.

Lang, Kenneth R. *The Cambridge Guide to the Solar System*. Cambridge, England: Cambridge University Press, 2003. A concise yet comprehensive book, containing a wealth of information on the members of the solar system. Excellent for a wide range of readers.

Reedy, Francis. "The Tenth Planet." *Astronomy* 33, no. 11 (2005): 68-69. A good basic article for a general readership, describing the scientific controversy over the definition of what constitutes a planet.

Sipiera, P. P. "Dawn Mission." In *USA in Space*. 3d ed. Edited by Russell R. Tobias and David G. Fisher. Pasadena, Calif.: Salem Press, 2006. A concise yet comprehensive article describing the current space mission to the asteroids Ceres and Vesta. Suitable for a wide range of readers.

COMET HALLEY

Category: Small Bodies

Halley's comet is the brightest, most famous of the known periodic comets. Definitive records of sightings go back more than two thousand years. The comet travels around the Sun roughly once every seventy-six years in a highly eccentric retrograde orbit inclined 20° to the ecliptic plane. Its orbital period has enabled many observers to see Halley's comet twice during their lifetimes.

Overview

For many years, the idea that comets were "dirty snowballs" has generally been accepted by astronomers. First proposed by Fred L. Whipple in 1950, this was one of a

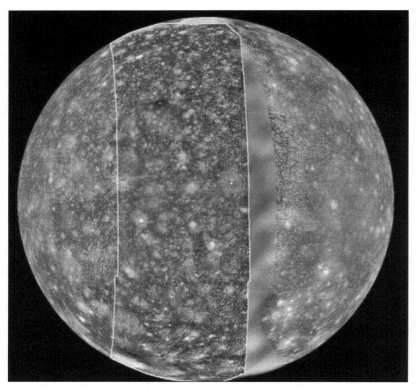

Comet Halley's nucleus. (NASA/GSFC)

objects were formed in the outer regions of the solar system, where the extremely low temperatures necessary for them to solidify prevailed. Giotto revealed that the nucleus of Halley's comet is a tiny, irregularly shaped chunk of ice coated by a layer of very dark material measuring some 15 kilometers long by 8 kilometers wide. This layer is thought to be composed of carbon-rich compounds and has a very low albedo, reflecting merely 4 percent of incident light. This low reflectivity makes the nucleus of Halley's comet one of the darkest objects known. However, various bright spots were seen on the nucleus. A hill-type feature was found near the terminator, along with one resembling a crater located near a line of vents. The vents seen on the nucleus appear to be fairly long-lived. Dust jets detected by the Russian Vega 1 and Vega 2 probes appear to have emanated from these vents, two of which were also identified by Giotto. Possibly some of the larger vents have survived successive perihelion passages.

Gas and dust that cause all the cometary activity seen (including the coma and tail) emanate from the nucleus via localized vents or fissures in the outer dust layer. These vents cover approximately 10 percent of the total surface area of the nucleus. They become active when exposed to the Sun and cease to expel material when plunged into darkness as the nucleus rotates. The force of these jets of material escaping from the nucleus plays an important role in the comet's motion around the Sun, affecting its orbital speed. Halley's comet was several days late in reaching perihelion during the last apparition in 1986, a result of the jetlike effects of the matter being expelled, as a consequence of Newton's second law of motion. The late arrival of Halley's comet was one of the factors examined by Swedish astronomer Hans Rickman, who attempted to calculate the mass of the nucleus from the amount of ejected material. Linking the ejection rate to the delay in perihelion, he judged the volume of the nucleus to be between 50 and 130 cubic kilometers. Measurements obtained through spacecraft imagery, however, revealed a volume closer to 500 cubic kilometers. The only conclusion was that the nucleus is markedly porous and far less

number of different ideas about the makeup of comets. The most popular idea was that they were "flying sandbanks," or collections of interstellar dust and gas accreted as the Sun and planets periodically passed through vast clouds of interstellar matter in their journey through the galaxy. The Sun's gravity then drew in the material that eventually collected to form individual bodies. This idea was popular during the first half of the twentieth century and was championed by British astronomers R. A. Lyttleton and Fred Hoyle. Since the middle of the nineteenth century, meteor streams have been associated with comets, and supporters of the "flying sandbank" model of cometary nuclei suggested that the particles within meteor streams arose from material escaping from comets as they moved through the solar system.

It is now widely believed that cometary nuclei are composed of material that condensed from the solar nebula at the same time as did the Sun and its planets. The European Space Agency's (ESA's) and other spacecraft that intercepted and studied Halley's comet in March, 1986, detected copious amounts of carbon, nitrogen, and oxygen. The materials given off by the comet signify that these dense than first anticipated, with an average density of no more than a quarter that of ice. This porosity meshes with the belief that comets formed in the outer regions of the solar nebula, where material coming together would remain loosely bound rather than compacting.

The fact that the nucleus of Halley's comet rotates is not in doubt. What does remain unresolved is the period of rotation. Using photographs of the comet taken during its apparition in 1910, astronomers calculated the rotation period to be 2.2 days around an axis that was fairly well aligned with the poles of the comet's orbit around the Sun. Results obtained by the Giotto, Vega, and Japanese Suisei probes appeared to support this value. Ground-based observations carried out during 1986, however, indicated a rotation period of 7.4 days. This value was supported by other ground-based observations, together with results from the American Pioneer Venus orbiter, which examined Halley's comet when it neared perihelion. Controversy ensued over these differing values, although a possible explanation has been suggested. The nucleus of Halley's comet could actually display both periods of rotation; one being spinning around its axis and another the precession of the axis of rotation. The combination of rotation and precession is still contested by certain astronomers, to some extent because of the porosity of the nucleus. Any precessional properties would quickly disappear unless the nucleus were fairly rigid.

Comets give off copious amounts of gas and dust that spread out in tails across large areas of space. Investigation of this material can reveal much about the composition of cometary interiors. Many of the investigations carried out by the European, Japanese, and Soviet space probes were directed toward a survey of the material ejected by Halley's comet. These investigations were supplemented by observations both from ground-based astronomers and from the American Pioneer Venus, International Cometary Explorer (ICE), and International Ultraviolet Explorer (IUE) spacecraft. As is the case with the surface of Halley's nucleus, the dust thrown off by the comet was found to be very dark and may have emanated from the surface itself rather than the interior. Giotto and Vega carried out analyses of the dust. They found a mixture of different materials, including the lighter elements oxygen, hydrogen, nitrogen, and carbon, and the heavier elements silicon, iron, and magnesium. The amount of carbon found during the investigations coincides quite well with the observed abundance of this material elsewhere in the galaxy, indicating that comets are made of interstellar material.

More than three-quarters of the gas ejected from the nucleus was found to be water vapor, which also appears to constitute more than 80 percent of the nucleus. The rate of production varied during the interval the comet was examined by these space probes. Vega 2 found approximately 16 tons of water coming away from the nucleus during its flyby, while Vega 1 detected double that rate. These large changes are reflected in the fact that Comet Halley's brightness sometimes varied by a factor of two or three from night to night. The velocity of the ejected vapor was found to be between 0.8 and 1.4 kilometers per second. This was the first time that water had been positively identified in a comet, in spite of the fact that cometary nuclei were widely thought to consist of a mixture of dust and water ice. Carbon monoxide and carbon dioxide were also detected, although methane was not found at all. This is strange, in that either any methane which existed in the comet may have been altered chemically during the period since the formation of the comet or methane was lacking in the cloud of material from which the comet formed. If there is methane in Halley's nucleus, it must constitute a very tiny percentage of the total makeup.

Processes involved in the release of gas from the nucleus may have played a prominent role in the evolution of its surface. It has been suggested that, as a comet approaches the Sun after spending its time in temperatures of approximately 40 kelvins in the outer regions of the Sun's influence, warming effects of the Sun can cause the ice within the nucleus to expand. This would result in heat generation and release of trapped gas. Some of this gas may collect in pockets, which eventually explode, producing craterlike features similar to that imaged by Giotto.

Methods of Study

Halley's comet is unusual (though not unique) in that it is named for the astronomer who first calculated its orbital path rather than the person who discovered it. Edmond Halley observed a bright comet in 1682, the impression of this sighting staying with him and eventually expanding into a deeper interest in comets. In 1705, Halley began a study of a number of bright comets seen between 1337 and 1698. Using methods developed by Sir Isaac Newton, he carried out work on the orbital motions of some twenty-four comets seen during this period. He noticed from his results that there were many similarities between the orbits of the comets observed in 1531 and 1607 and the bright comet he had seen in 1682. The intervals between the sightings were also roughly identical at around seventy-six years. This led Halley to predict that these sightings were of the same comet, and that it would reappear in 1758.

Halley died in 1743, although astronomers began a search for the returning comet as the date forecast by Halley drew near. French astronomer and mathematician Alexis-Claude Clairaut, with the help of Joseph-Jérome de Lalande and Madame Nicole Lepaute, attempted to calculate its orbital path in more detail. Taking into account gravitational effects of Jupiter and Saturn, they calculated that the comet would reach perihelion on April 13, 1759, and published ephemerides (detailed star maps and charts) to help astronomers with their search. Many famous astronomers joined in, although it was the amateur astronomer Johann Georg Palitzsh from Dresden who first spotted the comet on Christmas Day, 1758. The reappearance was quickly confirmed, and the comet was named for Halley in honor of the fact that he had correctly predicted its return. Once a number of observations had been obtained, a revised orbit was calculated. It was found

The Adventurous Edmond Halley

(NASA)

In 1684, Edmond Halley was a young scientist who had already made a name for himself as a precocious astronomer: He was the first to observe that the Sun rotated on an axis, during a trip to St. Helena in the South Seas. In 1680, during his Grand Tour of Italy and France, he had observed the comet that would bear his name. He had produced star catalogs and tidal tables, and he was trying to determine why Kepler's laws worked the way they did. Then, in April, his father's disfigured corpse was discovered near a riverbank; he had been missing for more than a month. Edmond's attention was redirected toward a bitter battle with his stepmother over the family estate.

Four months later, Halley was visiting Isaac Newton, who had solved the problems with Kepler's laws but had "misplaced" the solutions, supposedly worked out when Cambridge had been shut down during the plague of 1665. Halley began a campaign of diplomacy to get the eccentric and overly sensitive Newton to publish his results before someone else (Robert Hooke) derived the inverse square law and beat him to it. This was the genesis of Newton's *Principia* of 1687, published at Halley's expense.

In the meantime, Halley was supporting himself as a clerk at the Royal Society and working on a diverse array of projects, from determining the causes of the biblical Flood (which he unorthodoxly and dangerously placed earlier than the accepted date of 4004 B.C.E.) to making the connection between barometric pressure and altitude above sea level. He even calculated the height of the atmosphere, at a remarkably accurate 45 miles. Motivated by his persistent lack of money,

Halley also designed various nautical instruments: a prototype diving bell, a device for measuring the path of a ship, and another device for measuring the rate of evaporation of seawater. He even prepared lifeexpectancy tables that became the basis for modern life insurance. Between 1696 and 1698, he became the deputy comptroller of the Royal Mint at Chester, a post offered him by Newton, who was then the warden of the Mint. Administration did not prove to be one of Halley's many talents, however, and Newton found himself having to defend his friend against the Lord's Commissioners.

In 1698, Halley set out on another expedition to the South Seas to study the magnetic variations of the Earth's compass. The journey was abandoned (with the ship's first lieutenant facing a court-martial on their return), but Halley tried again a year later with more success. He also went on a secret mission in 1701, about which little is known, traveling to France for the Admiralty on the pretext of yet another scientific expedition. In 1703, Halley became a member of the Council of the Royal Society in recognition of his work, and in the same year, he was appointed to the Savilian Chair of Geometry at Oxford, where he conducted his study of comets. It was around this time that he made the observation for which he became famous:

> Many considerations incline me to believe that the comet of 1531 observed by Apianus is the same as that observed by Kepler and Longomontanus in 1607 and which I observed in 1682. . . . I would venture confidently to predict its return, namely in the year 1758. If this occurs there will be no further reason to doubt that other comets ought to return also.

In 1719, on the death of John Flamsteed, Halley succeeded to the post of Astronomer Royal, a position he held until his death in 1742. The practicality and range of his interests made him a celebrity whose achievements far exceeded those for which he is remembered today. He did not live to see his comet, which was sighted on Christmas, 1758.

that Clairaut's calculated perihelion date was in error by thirty-two days. Scientists were at a loss to explain this error, although they did not know about the existence of the two giant planets Uranus and Neptune, which were not to be discovered until 1781 and 1846, respectively.

Since the 1758 appearance, Halley's comet has been seen on three occasions: in 1835, in 1910, and in 1985-1986. Times of previous visits of the comet have been calculated by taking into account the gravitational effects of other bodies of the solar system and plotting the comet's orbital course backward through time. Dates calculated for previous apparitions have been substantiated by checking against ancient astronomical records, primarily those of Chinese astronomers. The first definite appearance of Halley's comet took place in 240 B.C.E., although the 12 B.C.E. appearance is the first about which detailed information is available. The most famous return was that of 1066, which was interpreted as a bad omen by the Saxons and, in particular, by Harold, the last of the Saxon kings. William of Normandy, who viewed the apparition as a good sign, invaded England, following which Harold died at the Battle of Hastings in October of that year. The Bayeux Tapestry depicts the comet suspended above Harold, who is seen tottering on his throne as his courtiers look on in awe and terror.

The 1531 appearance is important because it was one of two apparitions studied by Halley (the other being that of 1607) prior to his two deductions: that these historical sightings were of the same object, and that the comet is a regular visitor to this region of the solar system. A comprehensive set of observations of the 1531 appearance was made by astronomer Peter Apian, who published his results in 1540. The 1607 appearance was observed and recorded by many astronomers, including Johannes Kepler. This was the last apparition of Halley's comet before the introduction of the telescope.

After the comet's reappearance in 1758 and the discovery of Uranus in 1781, astronomers were able to plot its orbit with even greater accuracy. Long before its scheduled return in 1835, many attempts were made to calculate the expected date of perihelion passage. The consensus was that Halley's comet would pass closest to the Sun in November, 1835. The search for the returning comet started as early as December, 1834, almost a year before it was due to sweep through the inner solar system. The first sighting was not made, however, until August 6, 1835, by Father Dumouchel and Francisco di Vico at the Collegio Romano Observatory. Confirmation came via Friedrich Georg Wilhelm von Struve, who saw the comet on August 21. Perihelion occurred on November 16.

Prominent among the astronomers who studied the comet during the 1835 apparition was Sir John Frederick Herschel, who was then based at a temporary observatory near Cape Town, South Africa. He was in the process of completing the sky survey started by his father, Sir William Herschel, and had moved to South Africa in order to survey the southern stars that were visually inaccessible from England. John Herschel made his first attempt to locate the comet in late January, 1835, although he did not see it until October 28. The 1835 apparition was remarkable in that much activity was seen to occur in the comet. Prior to its temporary disappearance in the Sun's rays as it rounded the Sun, a number of changes were observed in the tail. These disturbances continued after its reappearance. The tail was seen to vary noticeably in length. The head also altered in appearance, at times appearing almost as a point of light, while at others taking on a nebulous form. It was noticed that the coma expanded while undergoing a reduction in brightness, eventually becoming so dim that it merged into the surrounding darkness. Herschel's final observation of Halley's comet in mid-May, 1836, was the last that any astronomer made until the 1910 return. All data scientists have about the 1835 apparition are in the form of sketches and visual descriptions. Photography had not yet made an impact on astronomy, although the appearance in 1910, through the use of the camera, provided the most comprehensive and detailed study of Halley's comet up to that time.

The third predicted return in 1910 was awaited eagerly by astronomers all over the world. The interval between the 1835 and 1910 visits had been littered with numerous bright comets, notable among which were the Great Comet of 1843, Donati's comet of 1858, and the Great September Comet of 1882. The latter is particularly significant in that it was the subject of the first successful attempt to photograph a comet. A good image was obtained by Sir David Gill in South Africa. Observation of Comet Morehouse in 1908 demonstrated that a series of photographs was an ideal means of monitoring cometary structural changes. (Comet Morehouse itself underwent a number of prominent changes that, coupled with the fact that Halley's comet had suffered in a similar fashion three-quarters of a century before, whetted the appetites of astronomers who were gearing up for the forthcoming apparition.) The prolonged period of cometary activity following its last visit had allowed astronomers to perfect their observing techniques and paved the way for observations of the return of Halley's comet.

The comet had passed aphelion in 1872, after which it once more began its long journey toward the inner solar system. The first astronomer to detect the returning visitor was astrophysics professor Max Wolf at Heidelberg, Germany. A photographic plate was exposed on the night of September 11-12 and recorded the comet close to its expected position. It did not become visible to the naked eye until well into 1910. Prior to this, another bright comet made an unexpected appearance. The Great Daylight Comet was first spotted by diamond miners in Transvaal, South Africa, in the early morning sky on January 13, 1910. Confirmation of the discovery was made four days later, and news of this spectacular discovery was distributed to the world's observatories. Unlike Halley's comet, which was to appear later that year, the Great Daylight Comet became a brilliant evening object for observers in the Northern Hemisphere. Its tail attained a maximum length of 30° or more by the end of January. The comet became so bright that it was visible to the naked eye even in broad daylight (hence its name).

The Great Daylight Comet was widely mistaken for Halley's comet by many people who had been expecting its return at about this time, although Halley's comet did not put on as grand a show. Bad weather together with the fact that a full moon occurred at what should have been the best time for observation meant that astronomers north of the equator were disappointed. Yet, even working against these odds, they did obtain many useful photographs and were able to study the comet spectroscopically. The best results, however, were obtained from observatories in the Southern Hemisphere, notably at Santiago in Chile. From mid-April to mid-May, 1910, Halley's comet was in the same area of the morning sky as Venus, the two objects together forming a marvelous visual spectacle in the constellation of Pisces. Much activity was noted in both the nucleus and the tail of the comet. Sequences of photographs showed marked changes in the head, including material being ejected from the nucleus and halos expanding out from the nucleus. The tail also underwent violent changes, with material

being seen to condense in various regions. On April 21, the day following perihelion, the previously smooth northern edge of the tail became irregular and distorted. Material seemed to be thrown out in various directions, and parts of the tail seemed to be ejected into space, an event clearly visible on photographs obtained at the time. For some days following perihelion, a jet of material from the nucleus seemed to be refueling the northern section of the tail. Once this activity ceased, the tail's southern section increased in brightness. A few weeks after perihelion, the two types of cometary tail appeared, a straight and distinct gas tail contrasting with the fainter, more diffuse and curved dust tail. Halley's comet passed between the Sun and Earth on May 18, although in spite of many attempted observations, no trace of the nucleus could be seen as the comet transited the solar disk. This proved that the nucleus must be tiny and the gas around it very tenuous. During this time, it was thought that the Earth may pass through the tail, although there is no evidence that this actually occurred. The pronounced curve of the tail seems to have taken it away from the Earth, preventing a passage of the planet through it. The closest approach of the comet to Earth was on May 20, when the distance between the two bodies was 21 million kilometers. For a time afterward, the comet became a prominent evening object for American observers, and many useful results

Launched in 1985, the Giotto space probe passed by Comet Halley's nucleus in 1986. (European Space Agency)

were obtained by astronomers at Lick Observatory and Mount Wilson Observatory in California. A number of changes in the comet's structure were seen, and many spectroscopic observations were taken. These showed the presence of a large number of different molecules in the comet, and helped astronomers to understand more clearly its chemical constitution.

As the comet started on its journey back to the outer regions of the solar system, it grew steadily fainter. It was last seen when beyond the orbit of Jupiter in a photograph taken on June 15, 1915, on its way toward aphelion in 1948. The next return would be accompanied by an unprecedented campaign by astronomers and space scientists to expand their understanding of comets in general, and Halley's comet in particular.

The return of 1985-1986, the most recent to date, provided astronomers with their best chance yet of exploring a comet. Unlike other bright comets, many of which appear suddenly, the orbital path of Halley's comet is known with both great precision and great accuracy. Therefore, it was possible to plan missions by robotic space probes to rendezvous with the comet during its last return. For a comet rendezvous mission, the position of the comet at time of interception must be known well in advance, as was the case with Halley's comet. In all, five space probes were sent to examine the comet.

Two of these were the Soviet Vega probes, launched in December, 1984, to release balloons into the Venusian atmosphere. Along their way, the probes encountered Halley's comet on March 6 and March 9, 1986, at distances of 8,890 kilometers and 8,030 kilometers, respectively. Among the equipment they carried were cameras, infrared spectrometers, and dust-impact detectors.

The two Japanese probes carried out their investigations from greater distances. Sakigake, launched in January, 1985, flew by the comet on March 11, 1986, at a distance of 6.9 million kilometers. Its primary purpose was to investigate the interaction between the solar wind and the comet at a large distance from the comet. One of the main aims of Suisei, launched in August, 1985, was to investigate the growth and decay of the hydrogen corona. Suisei flew past the comet on March 8, 1986, at a distance of 151,000 kilometers.

By far the most ambitious, and most successful, of the probes dispatched to Halley's comet was the European Giotto, named in honor of the Italian painter Giotto di Bondone. It launched toward the comet on July 2, 1985. Giotto was cylindrical in shape, with a length of 2.85 meters and a diameter of 1.86 meters. Its payload included numerous dust-impact detectors, a camera for imaging the nucleus and inner coma of Halley's comet, and a photopolarimeter for measuring the brightness of the coma. Giotto flew within 610 kilometers of the nucleus on March 14, 1986, at a speed of more than 65 kilometers per second. Data collected by Giotto were immediately transmitted back to Earth via a special high-gain antenna mounted on the end of the space probe facing away from the comet. Information was received back on Earth by the 64-meter antenna at the Parkes ground station in Australia. At the opposite end, Giotto was equipped with a special shield to protect it from impacts by dust particles during its passage through the comet's halo.

Exploration of Halley's comet by space probes was a truly international effort, the images and measurements obtained by the Soviet Vega craft helping scientists to target Giotto precisely. From Earth, the nucleus of a comet is hidden from view by the material surrounding it. Not until the Vega images were received was its position established and the subsequent trajectory of Giotto determined. During the close encounter, all instruments performed well, although disaster struck immediately before closest approach to the nucleus. A dust particle weighing merely one gram impacted Giotto. This temporarily knocked the spacecraft and its antenna out of alignment with Earth, and for thirty tense minutes contact was lost. The problem was rectified, and contact was reestablished. After the encounter, it was found that approximately half of the scientific equipment had suffered damage, although scientists were able to redirect the craft and put it on a course back to Earth. Tests carried out by the European Space Agency in 1989 paved the way for reactivation of Giotto, which set up a pass within 22,000 kilometers of the Earth and placed Giotto in a new orbit that allow it to intercept another comet. On July 10, 1992, Giotto flew close to Comet Grigg-Skjellerup, at a point just twelve days in advance of the comet's closest passage to the Sun, a time when its activity was approaching maximum.

Context

Although study of Halley's comet has taught scientists much about comets in general, there still remains much to learn about these ghostly visitors. Halley's comet provides a chance to investigate the origins of the solar system. Cometary explorations by space probes could include rendezvous missions during which a probe would position itself close to a cometary nucleus for a prolonged period and perhaps send a lander to the surface of the nucleus. The possibilities of such a mission were being examined by the National Aeronautics and Space Administration (NASA) at the time of Halley's 1986 visit. Known as Comet Rendezvous and Asteroid Flyby (CRAF), this mission would have enabled scientists to undertake close-up exploration of both asteroids and comets. Unfortunately, budget cuts led to the cancellation of CRAF. Sample return missions, by which scientists can examine at first hand material plucked from the heart of a comet, also remain a possibility. Some astronomers and scientists hope for a mission that will carry a human crew to Halley's comet during its next apparition, in 2061.

More realistically in the meantime, NASA was able to launch its Deep Space 1 probe and demonstrate the capability of an ion propulsion system to drive a spacecraft to effect rendezvous with an asteroid and a comet. On September 22, 2001, Deep Space 1 flew within 2,200 kilometers of Comet Borrelly, performing measurements and taking high-resolution images. The Deep Impact mission slammed a copper impactor into the Comet Tempel 1 on July 4, 2005, to expel surface material and excavate a crater on the comet's nucleus. The flyby portion of the Deep Impact spacecraft observed the collision of its impactor and analyzed material thrown up from the formation of an impact crater on the cometary nucleus. Then in January, 2006, the Stardust mission returned samples to Earth released from Comet Wild 2; those samples were collected at a distance of 240 kilometers from the comet's nucleus. ESA launched the Rosetta spacecraft in 2004 and set it on a trajectory toward an encounter with the comet 67P/Churyumov-Gerasimenko in May, 2014. Rosetta is designed to orbit the comet and later release a small lander named Philae to touch down on the comet's nucleus and perform in situ analyses of surface materials.

Brian Jones

Further Reading

Beatty, J. Kelly, Carolyn Collins Petersen, and Andrew Chaikin, eds. *The New Solar System*. 4th ed. Cambridge, Mass.: Sky, 1999. Filled with color diagrams and photographs, this popular work covers solar system astronomy and planetary exploration through the Mars Pathfinder and Galileo missions. Accessible to the astronomy enthusiast. Provokes excitement in the general reader, who gains an explanation of the need for greater understanding of the universe.

Faure, Gunter, and Teresa M. Mensing. *Introduction to Planetary Science: The Geological Perspective*. New York: Springer, 2007. Designed for college students majoring in Earth sciences, this textbook provides an application of general principles and subject material to bodies throughout the solar system. Excellent for learning comparative planetology.

Gingerich, Owen. "Newton, Halley, and the Comet." *Sky and Telescope* 71 (March, 1986): 230-232. Provides background information on Sir Isaac Newton and Edmond Halley. Describes how Halley used Newton's work with gravitation to draw comparisons between the orbital motions of comets seen in 1531, 1607, and 1682, which led to his conclusion that they were all sightings of the same object and his predict of its return in 1758. Suitable for the general reader.

Grewing, M., F. Praderie, and R. Reinhard, eds. *Exploration of Halley's Comet*. New York: Springer, 1989. A technical review of the information garnered by the return of Halley's comet in 1985-1986.

Harpur, Brian, and Laurence Anslow. *The Official Halley's Comet Project Book*. London: Hodder and Stoughton, 1985. A comprehensive guide to knowledge of Halley's comet prior to its exploration by space probe. As well as a general description of comets, the book contains details of Edmond Halley and his work, many facts relating to Halley's comet and its previous appearances, and a detailed description of the 1910 apparition of the comet. Includes a discussion on the pronunciation of Halley's name and a collection of poems written about the comet in 1910. A useful book for the general reader, containing many items not printed elsewhere.

McBride, Neil, and Iain Gilmour, eds. *An Introduction to the Solar System*. Cambridge, England: Cambridge University Press, 2004. A complete description of solar system astronomy suitable for an introductory college course, filled with supplemental learning aids and solved student exercises. Accessible to nonscientists as well. A Web site is available for educator support.

Sekanina, Zdenek, ed. *The Comet Halley Archive Summary Volume*. Pasadena, Calif.: Jet Propulsion Laboratory (International Halley Watch), California Institute of Technology, 1991. A collection of observations made by various observers of Halley's comet's return in 1985-1986.

Whipple, Fred L. "The Black Heart of Comet Halley." *Sky and Telescope* 73 (March, 1987): 242-245. An examination of the information received regarding the nucleus of Halley's comet and what it tells scientists. Comparisons are drawn between previous models of the structure of cometary nuclei and current knowledge. Suitable for the general reader.

_____. *The Mystery of Comets*. Washington, D.C.: Smithsonian Institution Press, 1985. Chapter 4, "Halley and His Comet," outlines the life and work of Edmond Halley and his involvement with cometary orbits. Chapter 5, "The Returns of Halley's Comet," describes the apparitions of Halley's comet from the earliest sightings to 1910; chapter 24, "Space Missions to Comets," is a description of the various space probes that intercepted Halley's comet during its return in 1985-1986. Suitable for the general reader.

COMET HYAKUTAKE IS DISCOVERED

Category: Small Bodies

The discovery of Comet Hyakutake, a comet of great brightness that passed very close to Earth, thrilled many and renewed discussions of the likelihood of a large body colliding with Earth at some point in the future.

Overview

Human beings have taken note of comets since ancient times. A comet is a celestial body that moves around the Sun; it consists of a central mass surrounded by a misty envelope that often forms a tail that streams away from the Sun. Comets vary considerably in the size of their central masses, and most comets are too small to be observed from Earth. Through the centuries, a few comets, such as Comet Halley, have attracted fame; these comets have been easily seen with the naked eye as they made repeated close passes by Earth.

In the twentieth century, improvements in technology led to the discovery of a number of comets, including Comet Hyakutake, which was discovered by Japanese amateur astronomer Yuji Hyakutake (comets are typically named for their discoverers). As a fifteen-year-old in Fukuoka, Japan, in 1965, Hyakutake saw Comet Ikeya-Seki in the night sky, and the experience led him to a lifelong interest in astronomy. As an adult, Hyakutake, who worked as a photoengraver, moved to Kagoshima, Japan, because the area's isolation meant that light pollution was low, allowing for relatively clear astronomical observations.

Hyakutake sought to discover a comet with a far orbit. To do so, he traveled to a rural mountaintop about

ten miles from his home to get a better view of the night sky. For four nights a month beginning in July, 1995, Hyakutake scanned the sky from 2:00 a.m. to 5:00 a.m. using only high-powered field binoculars with six-inch lenses. On December 26, he stayed a bit longer than usual and discovered a comet (later designated C/1995 Y1) at 5:40 a.m. This comet, which was not especially bright, attracted little attention outside the astronomy community. It could not be seen with the naked eye, but it was bright enough to be seen well with the use of a small telescope.

At 4:50 a.m. on January 30, 1996, Hyakutake discovered his second comet. He had returned to the mountain to take photographs of the first comet, but clouds in the comet's path foiled his plan. As he scanned the sky to find a clear spot, he saw a comet, but logic dictated that it could not be the same one he had seen in December because it was in almost the same location as the earlier sighting. Hyakutake realized that he had found a second body. He took photographs of the comet with a telephoto lens, developed the pictures, and, at 11:00 a.m., notified the National Astronomical Observatory in Tokyo of his discovery. Independent observations confirmed Hyakutake's find later that day. On February 27, 1996, Terry Lovejoy of Australia made the first naked-eye sighting of the comet, which had been designated C/1996 B2 or Comet Hyakutake.

With the discovery of Comet Hyakutake, sky watchers hoped for a truly big and active comet, but another spectacular comet along the lines of Comet Halley seemed unlikely. However, Hyakutake soon appeared to be brighter than even the original optimistic forecast. It produced roughly as much water vapor as Halley does at a comparable distance from the Sun as it approached Earth. To many experienced observers, Hyakutake qualified as a monster of a comet.

Comet Hyakutake remained visible with the naked eye for about one hundred days after February 27. On March 23, 1996, luminous knots of material—possibly small pieces of the comet's nucleus—were first observed moving back from its golden, starlike center. The head of the comet had an aquamarine hue. The apparent proportions of the comet's features were unlike anything the late twentieth century generation of observers had ever experienced, so they often found it difficult to interpret what they saw.

On March 23, the comet passed directly overhead the United States as seen from near 40 degrees north latitude. The comet passed perigee (the point at which it was closest to Earth) on March 25 at a distance of 9.3 million miles. Radar signals that bounced off Hyakutake's nucleus on that date indicated that the comet was surprisingly small, only about 0.6 to 1.9 miles in diameter. Ions disconnected from the tail from March 24 to March 26, and between April 8 and April 16 great tail lengths were again seen. The tail was detected out to about 60 degrees and possibly even 80 degrees or more. (A gas tail of 20 degrees is considered long.) Gas tails are more difficult to observe than dust tails because the eye is much less sensitive to the wavelengths of light emitted by gas tails than to the light reflected by dust tails. The greatest angular lengths of Hyakutake's tail were equivalent to about 20 million miles of gas tail.

By April 16, the comet's nucleus became less active and less bright, as noted in both the light curve and spectrographic observations. Perihelion passage (the point at which the comet was closest to the Sun) took place on May 1, 1997, at which time there was a distance of 21.4 million miles between the comet and the Sun. Lovejoy made the last naked-eye sighting of Comet Hyakutake in late June, 1996. It is not expected to be visible with the naked eye from Earth again for at least 72,000 years.

Yuji Hyakutake spent the remainder of his life searching fruitlessly for another comet. He died of an aneurysm at the age of fifty-one. In addition to the comet he discovered, Asteroid 7291 Hyakutake was named in his honor.

Significance

In 1996, Comet Hyakutake became the brightest comet to pass near Earth in more than four hundred years. Before it appeared, perhaps only thousands of living people in all the world had managed to get a good look at a great comet (that is, a comet of particularly great brightness). Hyakutake did not come as close to Earth as 1983's Comet IRAS-Araki-Alcock, but it captured a far greater amount of public attention. Part of this attention came from increasing fears that a close pass or collision with a comet could have disastrous consequences for the planet. A number of prominent scientists had speculated that a comet strike was responsible for environmental changes that led to the extinction of the dinosaurs, and many people had become concerned that such an event in the future would have similar consequences for humankind.

Although Comet Hyakutake thrilled many people who were interested in comets, the discovery had comparatively little significance within the scientific community. Part of the reason that most recent discoveries of comets had been made by amateur astronomers was

that professional astronomers were focusing their attention on other subjects, such as greater understanding of the other planets in our solar system. It is likely, however, that just as Yuji Hyakutake was inspired to study the skies by Comet Ikeya-Seki in 1965, future astronomers were inspired by Comet Hyakutake.

Caryn E. Neumann

Further Reading

Burnham, Robert. *Great Comets*. New York: Cambridge University Press, 2000. Lavishly illustrated volume provides a brief introduction to the history, nature, and beauty of comets. Includes index.

Crovisier, Jacques, and Thérèse Encrenaz. *Comet Science: The Study of Remnants from the Birth of the Solar System*. Translated by Stephen Lyle. New York: Cambridge University Press, 2000. Presents an overview of the forces and processes involved in the origin and evolution of comets. Intended for readers with strong background in science. Includes glossary, bibliography, and index.

Schaaf, Fred. *Comet of the Century: From Halley to Hale-Bopp*. New York: Copernicus Springer-Verlag, 1997. Combines solid history on comets and discussion of comet science with a personal account of the author's fascination with comets. Includes maps, photographs, and index.

COMETS

Category: Small Bodies

A comet is a minor body composed mostly of frozen ices typically embedded with solids. Comets revolve about the Sun in highly elliptical orbits. The Oort Cloud is a vast cloud of cometary bodies that extends billions of kilometers out from the Sun.

Overview

Comets are familiar to nearly everyone as majestic, star-like objects with long tails stretching across a wide band of the sky. The most famous comet, Halley's comet, makes its periodic return to the night skies every seventy-five years. The word "comet" is derived from a Greek word meaning "long-haired." Comets were greatly feared before the twentieth century as bad omens. Since then, they have been identified and cataloged as objects that come into the inner solar system from deep space. Most of them occupy orbits that carry them far away from the Sun. Many comets make only a single approach to the Sun and then never return again, while others exist in stable, but highly elliptical, orbits that allow them to return after an extended period of time.

One of the first theories advanced to explain the makeup of comets was proposed by astronomer Fred L. Whipple. Whipple suggested that comets were dirty snowballs, essentially bodies of water ice that incorporate dust and perhaps volatives other than water. This remained the primary theory through the first four decades of the space age. Only when spacecraft began visiting comets could it be put to the test.

Until the space age, comets were studied only in visible light through optical telescopic images. The first comet to be studied using Earth-orbital instruments, which permitted observations in the ultraviolet as well as the visible, was the much-heralded Comet Kohoutek late in 1973 and early in 1974. Comet Kohoutek turned out to be something of a disappointment visually from Earth, but images and data collected by the orbiting Skylab 4 astronauts advanced the understanding of comets.

In 1986, the European space probe Giotto passed about 600 kilometers from Halley's comet as the comet made its close approach to the Sun. The probe verified existing theories that comets are made up of ices covered by black dust or soil. In other words, the spacecraft confirmed the dirty snowball model, at least for this comet. Using data taken by the spacecraft, scientists determined that the dust is composed of carbon, hydrogen, oxygen, and nitrogen. Other metals have also been discovered in comets, such as iron, calcium, nickel, potassium, copper, and silicon. Halley's comet was one of the darkest objects ever seen in the solar system; it has virtually no albedo. Only one other major body in the solar system, Saturn's satellite Iapetus, is known to be this low in albedo.

As a comet approaches the Sun, it absorbs solar radiation and becomes warmer. The main body of the comet is called the nucleus. As the nucleus warms, ices beneath the comet's soil evaporate. Because the comet

A comet streaks across the night sky. (NASA)

has no atmosphere, evaporated substances, also called volatiles, escape into the vacuum of space. This gaseous envelope that surrounds the comet is called the coma. As the coma grows, it forms a plume of vapor that carries away some of the comet's surface dust as well. This mixture of evaporated volatiles and dust is carried away from the comet by the solar wind, is ionized by high-energy particles, and creates the spectacular tail of the comet. The comet's tail, glowing in the solar wind, can stream behind the comet for millions of kilometers. Cometary nuclei consist mostly of volatile ices and dust. That ice is nearly all water ice, but there is also evidence of ices composed of carbon dioxide and methane. More elementary compounds of nitrogen, oxygen, and carbon monoxide may exist as volatile ices.

Comets are typically small bodies. Halley's comet is an irregular potato-shaped object, 14 by 17 kilometers. In fact, some noted that images of Halley's comet captured during the Giotto mission suggested that the famous comet resembled the cartoon character Felix the Cat. The largest known comet is Chiron, which is estimated to be approximately 200 kilometers in diameter. Comets are thought to have formed as the solar system evolved. Comets were accreted out of material at the outer edge of the solar nebula that ultimately condensed to become the Sun and planets. Because cometary material was fashioned at the outer edge of the solar system, the Sun did not evaporate comets' volatiles. At the same time, the giant planets of the solar system formed at what would become the outer orbits of the solar system. These massive planets encountered the newly formed comets, and the comets

Facts About Selected Comets

Name (no)	Period (yrs)	Perihelion Date	Perihelion Distance	Distance from, Sun (AU)
Borrelly (19P)	6.86	2001-09-14	1.358	3.59
Chiron (95P)	50.7,	1996-02-14	8.460	13.7,
Crommelin (27P)	27.89	1984-09-01	0.743	9.20
d'Arrest (6P)	6.51	2008-08-01	1.346	3.49
Encke (2P)	3.30	2003-12-28	0.340	2.21
Giacobini-Zinner (21P)	6.52	1998-11-21	0.996	3.52
Grigg-Skjellerup (26P)	5.09	1992-07-22	0.989	2.96
Hale-Bopp	4,000, , ,	1997-03-31	0.914	250, ,
Halley (1P)	76.1,	1986-02-09	0.587	17.94
Honda-Mrkos-Pajdusakova (45P)	5.29	1995-12-25	0.581	3.02
Hyakutak	~40,000, , , ,	1996-05-01	0.230	~1,165, , ,
Kohoutek (75P)	6.24	1973-12-28	1.571	3.4
Schwassmann-Wachmann 3 (73P)	5.35	2006-06-02	0.933	3.06
Tempel 1 (9P)	5.51	2005-07-07	1.497	3.12
Tempel-Tuttle (55P)	32.92	1998-02-28	0.982	10.33
West-Kohoutek-Ikemura (76P)	6.46	2000-06-01	1.596	3.45
Wild 2 (81P)	6.39	2003-09-25	1.583	3.44
Wilson-Harrington (107P)	4.30	2001-03-24	1.000	2.64
Wirtanen (46P)	5.46	2013-10-21	1.063	3.12

Source: Data are from the National Aeronautics and Space Administration/Goddard Space Flight Center, National Space Science Data Center.

that were not engulfed by the giant planets were, over the first billion years, ejected into interstellar space by the planets' massive gravitational fields. Not all comets met that fate, however. Some were gently nudged into stable orbits closer to the Sun. Others were flung into the inner solar system, eventually impacting the inner planets. There are strong reasons to believe that Earth's oceans came from cometary ices delivered to the planet during the early era of bombardment, but that is not universally accepted.

What remained after billions of years of planetary encounters was an extraordinarily large cloud of comets extending outward from orbits beyond Pluto in all directions. A virtual spherically shaped cloud of comets surrounds the Sun at a distance from 1,000 to 100,000 astronomical units (AU). This cloud, which may contain as many as two trillion comets of all shapes and sizes, is called the Oort Cloud. It is named for the Dutch astronomer Jan Hendrik Oort, who first proposed its existence in 1950. The spherically shaped Oort Cloud is not the only source of comets in the solar system. There is a disk-shaped source of comets that extends from about 35 to 40 AU out from the Sun to about 1,000 AU. This source, the Kuiper Belt, was named for the astronomer Gerard Peter Kuiper, who theorized its possible existence in 1951. The disk-shaped Kuiper Belt blends with the spherical Oort Cloud at about 1,000 AU.

The Oort Cloud is the source for long-period comets, with orbital periods of greater than two hundred years. The Kuiper Belt is most likely the primary source for short-period comets, with orbital periods of less than two hundred years, such as Halley's comet. Comets have definite life spans, unlike planets. Each time a comet streaks in toward the Sun, volatile gases stream off the comet and form a beautiful cometary tail, while also depleting the comet's total mass. The comet melts away with each pass toward the Sun. When Halley's comet streamed past the Sun in 1986, the Giotto spacecraft measured a loss of 40 tons of mass per second from the comet. If the supply of comets were not

steadily replenished from deep space, they would have all been lost long ago.

The Sun is one among billions of stars in the Milky Way galaxy. In the relatively nearby region of the galaxy, there are hundreds of local stars, which are all revolving around the galactic center and are moving relative to one another. Because stars are so far apart on the average, the chance of one star colliding with another is quite low. However, the possibility of a local star passing near to or through the Oort Cloud (which extends up to 100,000 AU away from the Sun) is very high over millions of years. It is estimated that since the solar system formed, about five thousand stars have passed within 100,000 AU of the Sun. If an object as massive as another star passed close to the Oort Cloud, it could easily cause enough gravitational perturbations to direct comets in toward the Sun.

Since the Oort Cloud is spherical, long-period comets can appear to approach the Sun from any point in space. Short-period comets, originating from the Kuiper Belt, always appear to emanate from a band along the ecliptic

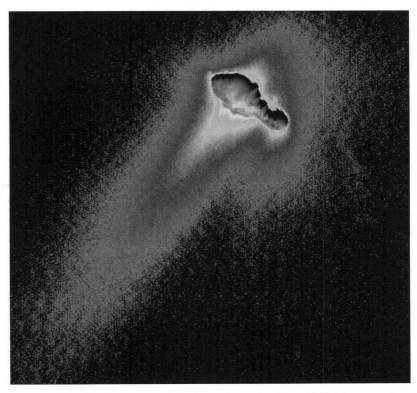

A composite of images captured by Deep Space 1, about 4,800 kilometers from Comet Borrelly, shows the comet's nucleus during the spacecraft's September, 2001, flyby. (NASA/JPL)

plane (the plane that contains the planetary orbits). After careful study of where comets actually originate, an analysis was made of their orbits. It has been discovered that there are areas of the sky that are richer in comets than others, and other areas that appear to be practically devoid of comets. Four different theories have been advanced to explain the source of these newly appearing comets. The first theory postulates that the passage of stars in or near the Oort Cloud may so affect the gravitational balance of comets that they are sent falling in toward the Sun. The second theory involves brown dwarfs, which are massive objects—about thirty times the mass of Jupiter—that are not quite planets and not quite stars. They do not have enough mass to create the conditions for thermonuclear ignition at their core. They predominantly radiate infrared energy, and they cannot be readily seen from Earth's surface. Current estimates approximate the number of brown dwarfs near the Sun to be sixty times greater than that of ordinary stars. A brown dwarf should pass through the Oort Cloud every 7 million years. Such an object would travel very slowly with respect to the Sun and would gravitationally release large swarms of comets into the solar system. These two stellar mechanisms, the action of either a passing star or a brown dwarf, are estimated to have been the source of about one-third of the observed comets.

According to the third theory, huge molecular clouds in interstellar space (much more massive than a single star) may pass at very large distances (tens of light-years) and may cause a release of comets through gentle perturbations of their orbits. The final theory for the source of newly appearing comets is galactic tidal action. Each galaxy has a gravitational field, which causes an attraction toward the midplane of the galaxy of all bodies (comets and stars). As these bodies orbit the galaxy, they are gravitationally influenced by one another. The galactic tide is the difference between the galactic forces acting on the Sun and the comet. Because the force of the galactic tide is very specific with respect to direction, it cannot act toward the poles of the Sun or toward the equator. Observations of cometary tracks confirm that comets from deep space do not seem to approach the Sun from these segments of the celestial sphere. This mechanism appears to explain the approach of the majority of all long-period comets entering the solar system from the Oort Cloud.

In the aftermath of the American decision to be the only spacefaring nation not to dispatch a spacecraft to investigate Halley's comet on its most recent appearance in the inner solar system, the National Aeronautics and Space Administration (NASA) proposed the Comet Rendezvous and Asteroid Flyby (CRAF). CRAF was a sister ship to the Cassini spacecraft. Because of budget cuts, NASA was able to save the Cassini mission, but CRAF was canceled. If it had been adopted, CRAF would have rendezvoused with Comet Kopff and remained in its vicinity for thirty-two months to observe variations in that comet's activity during different portions of its orbit. CRAF would also have dropped penetrometers into the comet to ascertain information about internal structure and make chemical analyses of surface materials.

Aspects of the ambitious CRAF concept were recycled into cheaper comet missions such as Deep Space 1, Stardust, and Deep Impact. Also, the European Space Agency (ESA) developed the Rosetta spacecraft to visit a comet.

Launched on October 24, 1998, the Deep Space 1 spacecraft began tests of an ion propulsion system and autonomous navigation system. Deep Space 1's targets were the asteroid Braille

The Spitzer Space Telescope captured this infrared image of the breakup of Comet 73P/Schwassman-Wachmann 3, which began to split into pieces in 1995. (NASA/JPL-Caltech)

and Comet Borrelly. Flying by the comet at a relatively close distance, Deep Space 1 captured images of a comet's nucleus that had higher resolutions than any of those captured by the probes that had visited Halley's comet. Comet Borrelly was shaped much like a bowling pin and displayed emission jets not distributed uniformly across its irregular nucleus. Deep Space 1 was not outfitted with debris shields, but it survived the close encounter nevertheless. In time Deep Space 1 ran out of propellant, but the mission showed that ion propulsion could be used on a spacecraft designed to visit multiple targets such as comets.

Stardust was launched on Feburary 7, 1999, and directed toward Comet Wild 2, where it opened up special sample collectors incorporating aerogel to capture both interplanetary and cometary dust. After the spacecraft's encounter with the comet on January 2, 2004, its sample collectors were sealed for a two-year journey back to Earth. On January 15, 2006, Stardust's sample collection unit safely reentered Earth's atmosphere and was recovered intact in Utah.

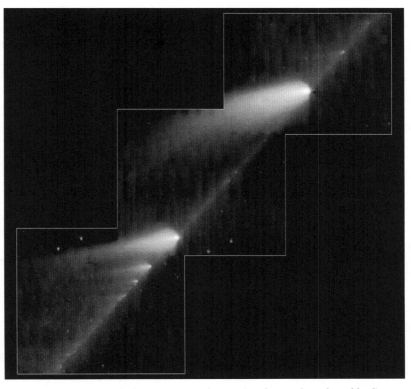

Debris from comet 73P/Schwassman-Wachmann 3, taken in the infrared by Spitzer. The debris passes near Earth every year and is expected to cause a significant meteor-shower display in the year 2022. (NASA/JPL-Caltech)

In June, 2008, researchers studying comet 26P/Grigg-Skjellerup material collected by the Stardust spacecraft announced that they had discovered new mineral grains. This mineral, named brownleeite after Donald Brownlee of the University of Washington, was a variety of manganese silicide not previously predicted by models of comets or the condensation of material from the early proto-Sun's nebula.

Deep Impact was designed as its name clearly suggests to fly to a comet and strike it, excavating material from deep below the surface. The spacecraft launched on January 12, 2005, and its onboard navigation steered the spacecraft toward the comet Tempel 1, released an impactor payload made largely of copper, and then veered out of the way to observe the resulting impact of the payload on the comet. The impactor was composed of copper, since that was an element not expected to be found naturally within the comet. The impact was observed by Deep Impact itself, as well as by the Hubble Space Telescope, the Chandra X-Ray Orbservatory, the Spitzer Space Telescope, the Swift spacecraft, and ESA's XMM-Newton observatory and Rosetta spacecraft. This coordinated effort permitted time-evolution studies of the plume and debris cloud created by the high-speed impact of the copper payload on Tempel 1.

Rosetta was launched on March 2, 2004. This was the European Space Agency's second attempt at a comet study and incorporated both a flyby craft and a lander named Philae. The mission was designed to rendezvous with the comet 67P/Churyumov-Gerasimenko in May, 2014, orbit it for many months while mapping the surface, and observe changes in the comet's activity as its distance to the Sun changed. Then the lander was scheduled to touch down on the comet on or about November, 2014, where it would secure itself to the surface in the comet's weak gravity field and then begin studies of chemical composition and physical characteristics of the comet's surface. The Rosetta mission was planned to continue through December, 2015.

Applications

The study of comets requires detailed knowledge of the composition of the outer regions of the solar system and the space between the last planet and 100,000 AU outward from the Sun. Comet studies also seek to understand complex gravitational interactions between bodies separated by wide distances and even gravitational interactions between tiny comets and the entire galaxy. Astronomers who study comets want to learn more about their makeup, their behavior when approaching the Sun, and the makeup and evolution of the early solar system.

New comets approaching the Sun for the first time have been held in deep freeze within the Oort Cloud and are thought to be composed of primordial material of the newly forming solar system. They have been tied up in the Oort Cloud for billions of years at temperatures barely above absolute zero. As they approach the Sun, their internal gases begin to stream away. Detailed study of an approaching comet's outgassing can inform planetary scientists about the composition of the early solar system. Comets and their approach have also hinted at the existence of the elusive brown dwarfs, thought to be one of the most common bodies of interstellar space. Because they are so dim, they are all but invisible from Earth. On the other hand, because brown dwarfs are thought to be so plentiful, the study of comets and their orbits may give the first real clues to the former's reality and abundance. The first serious studies of brown dwarfs came from observations made by the Spitzer Space Telescope, the final member of NASA's Great Observatory program. Spitzer detected brown dwarfs from their infrared emissions.

In the early 1980's the existence of galactic tidal action was merely speculation. Since then, careful study of comet orbits and their approaches to the inner solar system has favorably supported the theory of galactic tides. In the close approach of Halley's comet by robotic spacecraft in 1986, a wealth of information was recovered on the shape, behavior, and composition of comets. The existence of the Oort Cloud and the concept of gravitational interactions by passing objects in space have led to the theory of periodic comet showers. Such comet showers, separated by periods of tens of millions of years, may be responsible for certain mass extinctions on

Earth. These extinctions might be the result of a shower of comets originating from within the Oort Cloud, sent on their close approach to the Sun by the close passage of a star or brown dwarf to or through the Oort Cloud.

Samples of Comet Wild 2 were treated to many of the contamination safeguards used with the Apollo lunar rocks. Analyses of Stardust's captured comet material revealed some surprises. Tracks in the aerogel suggested solid materials larger than interstellar dust grains. Silicate crystals and other mineral crystals were found which required more than just mild heating, as would have been the case if the comet was largely composed of interstellar dust grains. This suggested that the theory of comet formation may need alteration. Inclusions of vanadium nitride, titanium, molybdenum, osmium, ruthenium, and tungsten were found, components that would have required high heating. Samples also contained organic materials more primitive than found in asteroidal material, compounds such as polycyclic aromatic hydrocarbons.

The collision of Deep Impact's copper payload was equivalent to five tons of TNT. Comet Tempel 1

Comet Tempel 1 in an image taken in 1998 by the Deep Impact flyby craft. (NASA/JPL-Caltech)

increased in brightness sixfold as a result of the event. As much as between 10 and 25 million kilograms of comet material was ejected as a crater formed as a result of the impact. Tempel 1 material was much finer than had been expected, being more akin to talcum powder than a sandy grain. Data ruled out a loose aggregate or highly porous model of the comet's structure. Indeed, rather unlike a dirty snowball as proposed by Whipple, Comet Tempel 1 was more like an icy dirt ball. Seen in the ejected materials in addition to volatiles were clays, carbonates, sodium, and crystalline silicates. After this encounter the flyby portion of the Deep Impact spacecraft was redirected to a planned encounter with Comet Hartley 2 in 2010.

Rosetta holds the potential for in situ analyses of comet material if the Philae lander successfully touches down safely, and can perform its chemical and physical tests. This information will greatly assist in determining how comets form and how their comas and tails develop as they travel into the inner solar system.

Context

Humankind has always looked to the heavens in awe and wonder, and sometimes in fear. Perhaps no other astronomical phenomenon except a total solar eclipse has historically evoked as much fear as comets. When the specter of fear is removed, however, they emerge as strikingly beautiful objects in the sky. It was once believed that if Earth passed through the tail of a comet, its inhabitants would die; this theory has been discredited. Comets are messengers from a time long past. Most are chunks of dirty ice, locked away in the Oort Cloud for billions of years.

Comets have been used to judge vast distances, evaluate the early composition of the solar system, and even test the idea that the gravity of the entire Galaxy can make a difference to the smallest of objects in space. Comets have been used as yardsticks to evaluate what may be the most common type of star in the galaxy, the brown dwarf—which ironically is one that is difficult to observe, even in the infrared. Comets have been called dirty snowballs. Halley's comet was so black that it was the darkest object ever seen in space. Yet, from these dirty specks of ice, planetary scientists have witnessed some of the most spectacular light shows. Ultimately, comets may also generate clues to some of the most fundamental secrets about the solar system and planets. From these tiny messengers, planetary scientists may unlock and examine pristine elements from the formation of the solar system.

Debris from comets provides the material that Earth passes through when annual meteor showers occur. For example, the Orionid meteor shower is leftover material from Halley's comet, the Leonids meteor shower is associated with Comet Tempel-Tuttle, and the Perseid meteor shower is material from Comet Swift-Tuttle.

Historically comets have come full circle from being seen as omens in the heavens to be feared, to celestial objects evoking a sense of wonder, and to again being objects that should be feared if they come too close and perhaps even impact Earth. Comets represent a more troubling threat than do asteroids, as comets are usually discovered only when they come in past the orbit of Jupiter. As such there is insufficient time to mount any mitigating effort if a new comet is determined to make a close pass or actually impact the Earth. Have comets impacted the Earth in the past? The answer is believed to be almost certainly yes, and indeed many believe that the majority of Earth's water came from comets encountering the early Earth. The Siberian Tunguska event of 1908, itself a curiosity that has been explained by some (without any legitimate supporting evidence) as a nuclear explosion or even the impact of an unidentified flying object, is now believed to have been the result of a comet or asteroid impact, most likely an air burst explosion of the body. Although this theory is not yet confirmed, it points out the potential for devastation that an impacting comet represents.

A comet collision was observed in 1994 when the nearly two dozen pieces of the shattered Comet Shoemaker-Levy 9 smacked into Jupiter's upper atmosphere. These pieces created temporary changes in the gas giant's appearance, many of which were the size of the Earth, indicating that a tremendous amount of energy was involved in this series of collisions. The impacts were recorded by the Hubble Space Telescope and Galileo spacecraft; the incredible magnitude of the disruption came as a surprise to the scientific community.

It was believed since roughly 1950 that gravitational disruption of the Sun's Oort Cloud by a close passage of another star was responsible for swarms of comets heading into the inner solar system, resulting in bombardment of the planets. However, in late 2008 Hans Rickman of Sweden's Uppsala Astronomical Observatory reported in *Celestial Mechanics and Dynamical Astronomy* the results of an updated computer simulation of the Oort Cloud investigated by his research group. If correct, their model indicates that sporadic stellar encounters, while indeed important in generating fresh comets that head toward the inner solar system, is not the only mechanism for sending comets toward the planets. This model accounted for galactic gravitational tidal influences on the Oort Cloud and found that the threat from comets may be more constant than previously believed. If correct, the model reinforces the need to monitor the skies for incoming comets that might be headed our way.

Dennis Chamberland and David G. Fisher

Further Reading

Arny, Thomas T. *Explorations: An Introduction to Astronomy*. 3d ed. New York: McGraw-Hill, 2003. A general astronomy text for the nonscience reader. Includes an interactive CD-ROM and is updated with a Web site.

Benningfield, Damond. "Where Do Comets Come From?" *Astronomy* 18 (September, 1990): 28-36. A fine summary of comets, superbly illustrated and written for the general public. Addresses the question of the Oort Cloud and Kuiper Belt in detailed, scaled illustrations. Discusses possible linkage to the extinction of the dinosaurs and the latest satellite discoveries.

Brandt, John C., and Robert D. Chapman. *Introduction to Comets*. New York: Cambridge University Press, 2004. A text suitable for a planetary science course, this comprehensive work covers our knowledge of comets from early observations to telescopic investigations through spacecraft encounters of these mysterious and alluring bodies.

Hartmann, William K. *Moons and Planets*. 5th ed. Belmont, Calif.: Thomson Brooks/Cole, 2005. An updated version of a classic text that covers all aspects of planetary science. Well-explained material on asteroids and comets. Takes a comparative planetology approach.

Levy, David H. *The Quest for Comets: An Explosive Trail of Beauty and Danger*. New York: Plenum Press, 1994. Written by one of the codiscoverers of Comet Shoemaker-Levy 9, this book is for the general reader. It highlights the author's comet discovery program and the comet catastrophe theory.

Newburn, R. L., M. Neugebauer, and Jurgen H. Rahe, eds. *Comets in the Post-Halley Era*. New York: Springer, 2007. A collection of fifty papers compiled into a volume in Springer's Astrophysics and Space Science Library. Covers observational techniques, origin of comets, evolution of comets, and spacecraft data. Special attention is given to Comet Halley and other comets encountered by spacecraft.

Russell, Christopher T. *Deep Impact Mission: Looking Beneath the Surface of a Cometary Nucleus*. New York: Springer, 2005. A complete description of the Deep Impact mission to excavate and analyze material from Comet 9P/Tempel 1.

Sagan, Carl, and Ann Druyan. *Comet*. New York: Random House, 1985. This coffee-table book is a classic work of art, written by the most popular astronomer in the United States. Filled with beautiful color and historical black-and-white photographs and illustrations, it was the basis for the popular movie by the same name. For general audiences.

Schaaf, Fred. *Comet of the Century: From Halley to Hale-Bopp*. New York: Copernicus Books, 1997. Comet Hale-Bopp was a popular comet to observe, one that provided a better show than the most recent appearance of Halley's comet. For a general audience.

Thomas, Paul J., Roland D. Hicks, Christopher F. Chyba, and Christopher P. McKay. *Comets and the Origin and Evolution of Life*. 2d ed. New York: Springer, 2006. A collection of chapters written by experts in the field. This update of the first edition covers new understandings of Halley's comet and more recent spacecraft data. Provides insights into organic compounds found in comets, protostars, and interstellar clouds.

Verschuur, Gerrit L. *Impact! The Threat of Comets and Asteroids*. New York: Oxford University Press, 1997. Verschuur explains the change in thinking from uniformitarianism to catastrophism. Identifies the Chicxulub Crater with an impact event that led to the extinction of the dinosaurs. Warns of the potential for devastation that a comet impact on Earth would cause.

COMET SHOEMAKER-LEVY 9

Category: Small Bodies

The spectacular collision of comet Shoemaker-Levy 9 with Jupiter in July, 1994, provided valuable information about comets, Jupiter's atmosphere, and the Jovian role in diminishing potentially cataclysmic Earth-damaging debris in the inner solar system.

Overview

On the night of March 24, 1993, the husband-and-wife team of Eugene ("Gene") and Carolyn Shoemaker together with their colleague David Levy were using the Schmidt telescope at Mount Palomar Observatory in

California to take photographs in connection with a project designed to discover near-Earth celestial objects. Gene Shoemaker, who had recently retired from the U.S. Geological Survey's Astrogeology Research Program, which he had established, was an expert in Earth-orbit-crossing asteroids and comets. In their five years together, the Shoemaker-Levy team had already discovered eight comets, and they were pleased when one of their photographs of the sky near Jupiter revealed what Levy called "the strangest comet" he had ever seen. Its several tails and bat-shaped wings of dust reminded him of the American Stealth Bomber. They quickly realized that this fragmented comet was an important discovery, and, three days later, they made comet Shoemaker-Levy 9 (SL 9) public in a circular published by the International Astronomical Union (IAU). Following tradition, the comet was named after its discoverers, with the number indicating it was the ninth comet that this team had found. Its formal IAU name became D/1993F2, in which the prefix "D" indicated that it was a periodic comet that later "disappeared," 1993 was the year of discovery, the suffix "F" represented the half-month of discovery (F = March 16-31), and "2" meant it was the second discovery in that half-month.

Other observers, stimulated by this announcement, returned to photographs that they had taken before March 24 and confirmed the discovery of the Palomar team. Using these and other data, astronomers calculated the orbit of the comet, which, unlike all earlier comets, orbited Jupiter rather than the Sun. Its highly elliptical orbit had an apojove (the point farthest away from Jupiter) of nearly 50 million kilometers and a period of nearly two years. Later data helped refine this orbit and provided clues about the comet's early history. Like most comets, it had orbited the Sun, but several decades earlier it had been captured by Jupiter's gravity. In early July, 1992, it traveled so close to Jupiter, about 20,000 kilometers from the Jovian cloud tops, that the giant planet's powerful gravity broke the comet into twenty-one separate fragments, each of which collected a coma of dust.

Following the convention established for previous fragmented comets, the twenty-one discernible pieces were labeled with letters of the alphabet (excluding I and O), and so SL 9's fragments, which averaged a few kilometers in diameter, ran from A to W, with the brightest (and presumably the largest) piece called Q. With the orbit and fragments identified, astronomers soon realized that this piecemeal comet was on a collision course with Jupiter. For SL 9 there would be no escape from Jovian orbit to return to the Kuiper Belt, now believed to be the source of Jupiter Family (JF) comets; instead, SL 9 faced extinction within sixteen months of its discovery.

Because astronomers knew when the comet would collide with Jupiter, they had time to organize observatories around the world, including in such remote locations as Antarctica, to make preparations to collect data on this unprecedented event. Because the initial impact would take place on the side of Jupiter hidden from the Earth, several spacecraft would play important roles, particularly in the earliest observations. These spacecraft included Galileo, well on its way to study Jupiter; the

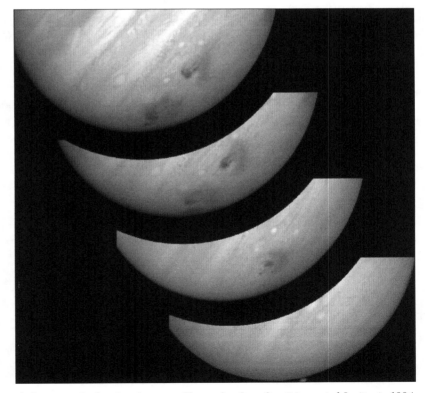

A photo mosaic showing the comet Shoemaker-Levy 9 as it impacted Jupiter in 1994. (JPL/NASA/STScI)

Hubble Space Telescope (HST); Ulysses, which had been studying solar poles; the Roentgen satellite (ROSAT), which had been surveying the sky for X-ray sources; and Voyager 2, which had been exploring the outer planets of the solar system.

As predicted, the comet's first fragments slammed into the Jovian atmosphere on July 16, 1994, at a speed of about 60 kilometers/second, or fast enough to traverse the United States in about a minute. When Jupiter's rotation made the crash site visible to terrestrial observers, thousands of telescopes could see the dark spots that had been created. Fragments of SL 9 continued to collide with Jupiter over the next 5.6 days. Because of the great excitement created by this astronomical event, and because the Earth had been interconnected by various computer networks, images, observations, scientific data, and personal impressions were rapidly transmitted all over the planet. Many others experienced the event through television or through the many stories in magazines and newspapers.

By the time the final fragment, W, struck Jupiter on July 22, many millions of Earthlings had shared, in some way or other, this unique interplanetary event.

Knowledge Gained

The prodigious wealth of information created by the SL 9 event had important implications for the understanding of comets, the Jovian atmosphere, and the future history of Earth. Astronomers knew, of course, that comets could be destroyed by collisions with the Sun, planets, and satellites, but the data from the fragmentation and collision of SL 9 with Jupiter revealed that its nucleus had been neither a solid body nor a loose agglomeration of materials but something in between. When the pieces hit Jupiter, spectroscopic analysis detected the presence of several elements absent from the Jovian atmosphere. These elements, which came from the comet, included such nonmetals as sulfur and silicon and such metals as iron, aluminum, magnesium, and even lithium, hitherto undetected in comets.

The remains of Shoemaker-Levy 9 emerge after the comet's impact with Jupiter caused it to break into twenty-one pieces. (JPL/D. Seal, edited by CXC/M. Weiss)

As expected, when the high-speed fragments of SL 9 penetrated the Jovian atmosphere, gigantic explosions and massive seismic waves resulted. Fireballs created temperatures in excess of 10,000 kelvins, which rapidly diminished to 2,000 kelvins. Collision-zone temperatures remained elevated for two weeks, but, astonishingly, smaller impact sites had higher temperatures than larger ones. A typical fireball spread from 15 to 100 kilometers in about 40 seconds, and some plumes extended to an altitude of 3,000 kilometers. These explosions also produced waves that sped across the planet at about 450 kilometers/second. These waves, which weakened in about two hours, posed a problem for astronomers. Disagreements developed about where they occurred (in the Jovian stratosphere or troposphere) and how they traveled (guided by a stable layer or generated by interlayer complexities). Just as spectroscopic analysis revealed some surprises about SL 9's chemical composition, so, too, certain elements and compounds were discovered for the first time on Jupiter: for example, diatomic sulfur and carbon disulfide. By contrast, astronomers had expected to find sulfur dioxide, but they did not.

Astronomers also used other parts of the electromagnetic spectrum to gather data on the collision. For example, radio emissions at a specific wavelength (21 centimeters) were indicative of synchrotron radiation, most likely caused by the collision's injecting very-high-speed electrons into the Jovian magnetosphere, which also experienced other changes after the impact. Since both Jupiter and comets were known to have water in their makeup, astronomers were surprised by the very small amounts of water that were detected. Perhaps the comet's fragments lost most of their water before the collision, or perhaps the comet's fragments were destroyed before they reached the planet's water layer.

For many people the highlight of the event was the creation of a series of dark spots that scarred Jupiter's southern hemisphere for several weeks. Similar to the Great Red Spot, if smaller in scale, these dark spots were the most enduring transient features ever seen on the planet, although some historians of science pointed out that, in 1790, Gian Domenico Cassini had reported unusual temporary marks on Jupiter's disk. If these had been due to a cometary collision, then SL 9's crash onto Jupiter would not have been the unique event that many touted it to be.

Context

Throughout its long history, the Earth has experienced steady and numerous collisions with interplanetary objects. The collision of the Shoemaker-Levy 9 comet with Jupiter provided

astronomers with valuable insights into how the collision affected Jupiter and, by analogy, how such comets may have affected other planets, including Earth. Gene Shoemaker estimated that comets had most likely caused about a fifth of the large impact craters on Earth. Linear crater chains have been photographed on Ganymede and Callisto, two of Jupiter's satellites, and these were probably due to cometary collisions. Some scientists have speculated that if SL 9 had collided with Earth instead of Jupiter, a cataclysmic destruction of life would have occurred. According to many scientists, 65 million years ago an asteroid or comet smashed into Central America, creating massive amounts of atmospheric pollutants that helped to bring about the extinction of the dinosaurs and many other forms of life.

In the history of life on Earth other mass extinctions have occurred, and some scientists associate these with periodic comet showers. Various theories have been put forward to explain these periodicities—for example, Nemesis, a companion star of our Sun, may create perturbations in the Oort Cloud that lead to these recurrent invasions of comets into the solar system. Jan Oort was the first to suggest that this large reservoir of icy bodies might be the source of very-long-period comets. However, SL 9's collision with Jupiter revealed something very significant. Because of Jupiter's powerful gravitational field, it attracts many asteroids, comets, and other interplanetary debris, resulting in fewer collisions of these objects with the inner planets, especially Earth. Some have even called Jupiter a "cosmic vacuum cleaner." On the other hand, estimates indicate that small comets collide with Jupiter about once a century, and comets comparable in size to SL 9 hit it about once per millennium. Comet Shoemaker-Levy 9 certainly expanded knowledge about the nature and properties of comets, as well as their interactions with other members of the solar system, but astronomers also realize that they need to learn much more before they will be able to make reliable predictions about some future comet's possibly devastating collision with Earth.

Robert J. Paradowski

Further Reading

Fernández, Julio Angel. *Comets: Nature, Dynamics, Origin, and Their Cosmological Relevance.* Dordrecht, Netherlands: Springer, 2005. Using advanced mathematics and celestial mechanics, the author analyzes the history, structure, and behavior of comets, including SL 9. Includes 483 references and an index.

Gehrels, Tom. "Collisions with Comets and Asteroids." *Scientific American* 274, no. 3 (March, 1996): 54-59. A discussion of the likelihood of cometary and asteroid impacts on Earth, designed for the scientifically inclined general audience.

Levy, David H. *Impact Jupiter: The Crash of Comet Shoemaker-Levy 9*. New York: Basic Books, 2003. Called the definitive memoir of SL 9, this book by the comet's codiscoverer provides for the general reader a lively account of the comet, its collision with Jupiter, and the knowledge gained from this event. Illustrated with color and black-and-white photographs. Chapter references and an index.

Noll, Keith S., Harold A. Weaver, and Paul D. Feldman, eds. *The Collision of Comet Shoemaker-Levy 9 and Jupiter*. Cambridge, England: Cambridge University Press, 1996. Contains fifteen reviews by experts involved with the SL 9 event, as well as many references to the primary literature. Index.

Shoemaker, Eugene M., P. R. Weissman, and C. S. Shoemaker. "The Flux of Periodic Comets Near Earth." In *Hazards Due to Comets and Asteroids*, edited by Tom Gehrels. Tucson: University of Arizona Press, 1994. This article explores the probability of a comet like SL 9 colliding with Earth.

Spencer, John R., and Jacqueline Mitton, eds. *The Great Comet Crash: The Collision of Comet Shoemaker-Levy 9 and Jupiter*. Cambridge, England: Cambridge University Press, 1995. A collection of articles by various scientists, including the Shoemakers. Analyzes, from various perspectives, the discovery, tracking, and crash of this comet into Jupiter, as well as discussions of the many discoveries this event stimulated. In the final selections, astronomers explore the possible effects of a comet like SL 9 colliding with Earth.

DWARF PLANETS

Categories: Planets and Planetology; Small Bodies

The discovery of many new bodies orbiting the Sun beyond the orbit of formerly outermost Neptune—including at least one larger than Pluto—created a crisis in astronomy. It became evident that a new definition was required to distinguish these objects from traditional planets. The term "dwarf planet" was introduced to include planetary objects smaller than planets but larger than asteroids, resulting in the demotion of Pluto from its status as a planet.

Overview

The concept of a planet has a long history, leading to a total of nine planets in the solar system until discoveries in the early twenty-first century led to new definitions that excluded Pluto. The word "planet" originates from a Greek word meaning "wanderer" and for centuries was applied to celestial objects that shifted positions relative to the "fixed" stars. In classical antiquity, seven such objects were identified and were associated with mythical gods: the Sun, the Moon, Mercury, Venus, Mars, Jupiter, and Saturn. The Latin names for the seven days of the week were based on these seven celestial deities. In Greek thought, the planets were believed to orbit the Earth along complex paths determined by a combination of circles.

During the scientific revolution of the sixteenth and seventeenth centuries, it was shown that five of the classical planets revolve around the Sun in elliptical orbits, along with the Earth-Moon system. Late in the eighteenth century, British astronomer William Herschel, aided by his sister Caroline, discovered Uranus, the first planet to be discovered with the aid of a telescope. Early in the nineteenth century, Sicilian astronomer Giuseppe Piazzi discovered what he thought was a new planet, smaller than Mercury and orbiting the Sun between Mars and Jupiter. He called it Ceres. However, when many smaller bodies with similar orbits were discovered in the next few decades, they were called asteroids, and Ceres was demoted from its status as a planet (though later reinstated as a "dwarf planet"). The asteroids are believed to be remnants from the formation of the solar system.

By the middle of the nineteenth century, investigations into slight deviations in the elliptical orbit of Uranus led to the discovery of Neptune by the German astronomer Johann Galle. Using Sir Isaac Newton's law of universal gravitation, astronomers were able to determine the masses of all but two of the eight known planets from the motions of their satellites, with Jupiter as the most massive, at 318 times the Earth's mass. Perceived deviations in the orbit of Neptune led to the discovery of Pluto in 1930 by American astronomer Clyde Tombaugh. Pluto's orbit differed from those of the other planets, with its large inclination from the ecliptic plane and its highly elliptical shape that brings it closer to the Sun than Neptune during some 20 years of its 248-year period. It was also found to be much smaller than the outer gas giant planets and to consist mostly of icy materials.

In 1977 Charles Kowal discovered a small, icy planetoid orbiting the Sun between Jupiter and Uranus, later named Chiron. In the 1990's several similar, cometlike

Sedna
800-1100 miles
in diameter

Quaoar
(800 miles)

Pluto
(1400 miles)

Moon
(2100 miles)

Earth
(8000 miles)

The dwarf planets Sedna and Pluto shown in size comparisons with other bodies. (NASA/JPL-Caltech/R. Hurt, SSC-Caltech)

objects were found between Jupiter and Neptune and are now called centaurs. Pluto's status as the ninth planet began to be suspect in 1978, when its satellite Charon was discovered and Pluto's mass was found to be only 0.2 percent of Earth's mass. That is much less than even Mercury, at 5.5 percent of the mass of the Earth. Pluto's mass was too small to have produced deviations in Neptune's orbit, which were then found to be negligible.

Then, in 1992 after a five-year search using digital cameras and computerized analysis, David Jewitt and Jane Luu of the Massachusetts Institute of Technology (MIT) discovered the first of many similar icy objects beyond Neptune in a region called the Kuiper Belt. Existence of such a region had been predicted by Dutch

American astronomer Gerard Kuiper. It is similar to the asteroid belt between Mars and Jupiter but about twenty times wider and populated by icy objects rather than the rocky and metallic bodies found in the asteroid belt. The Kuiper Belt extends from the orbit of Neptune, between 30 and about 55 astronomical units (AU), and is believed to contain thousands of objects larger than 100 kilometers in diameter.

More than 130 Kuiper Belt objects (KBOs) have been found with nearly the same 248-year period as Pluto at about 40 AU from the Sun. These "plutinos" complete their orbits twice during three orbits of Neptune, referred to as a 2:3 gravitational resonance. KBOs with other resonances, such as 3:5 and 4:7, are called cubewanos, and a few objects are found

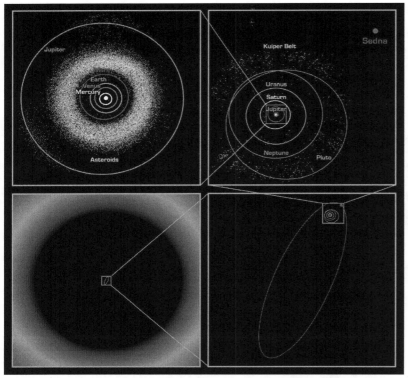

Four panels show the location of the dwarf planet Sedna. (NASA/JPL-Caltech/R. Hurt, SSC-Caltech)

The definition of a planet was placed on the agenda for the General Assembly of the International Astronomical Union (IAU) meeting in August of 2006. An initial draft proposal recommended that Pluto be retained as a planet and that Ceres, Charon, and Eris be added to the list of planets. This recommendation was made by astronomers who viewed both Pluto and its satellite Charon as planets in a double-planet system, since each body rotates about a point located between the two. After many objections, an alternate proposal was offered by the Uruguayan astronomer Julio Fernández, who suggested an intermediate category for objects like Pluto, which are large enough to be nearly round but too small to clear their orbits of other planetesimals. The IAU accepted this proposal, and by unanimous vote it was agreed to call these intermediate objects "dwarf planets," with smaller objects to be called "small solar-system bodies." By further vote, it was agreed that Pluto is a dwarf planet.

These definitions as voted in Resolution 5A by the Twenty-sixth General Assembly of the IAU are as follows:

The IAU therefore resolves that planets and other bodies in our Solar System, except satellites, be defined into three distinct categories in the following way:

(1) A "planet" is a celestial body that (a) is in orbit around the Sun (b) has sufficient mass for its self-gravity to overcome rigid body forces so that it assumes a hydrostatic equilibrium (nearly round) shape, and (c) has cleared the neighbourhood around its orbit.

(2) A "dwarf planet" is a celestial body that (a) is in orbit around the Sun (b) has sufficient mass for its self-gravity to overcome rigid body forces so that it assumes a hydrostatic equilibrium (nearly round) shape (c) has not cleared the neighbourhood around its orbit, and (d) is not a satellite.

(3) All other objects, except satellites, orbiting the Sun shall be referred to collectively as "small solar-system bodies."

beyond a 1:2 resonance at 55 AU and with 330-year periods. Some objects have been found beyond 55 AU but are believed to have been scattered from the Kuiper Belt into a region called the scattered disk containing scattered disk objects (SDOs). Planetesimal objects in these latter two regions (KBOs and SDOs) are called trans-Neptunian objects (TNOs).

Astronomers began to view Pluto as the largest member of the new class of plutinos, and some started to question its status as a planet. In 2003 a team from the California Institute of Technology (CalTech), working at Mount Palomar Observatory north of San Diego and led by Mike Brown, discovered an SDO at about 97 AU from the Sun, now called Eris. When a satellite was discovered in 2005, the mass of Eris was found to be 27 percent larger than that of Pluto, and a few astronomers began to refer to it as the tenth planet. Most astronomers, however, recognized that many TNOs might be larger than Pluto and that either they would also have to be classified as planets or Pluto would have to be reclassified to distinguish such objects from the traditional planets.

In three footnotes, this IAU resolution agreed that the eight planets are Mercury, Venus, Earth, Mars, Jupiter, Saturn, Uranus, and Neptune. It also agreed to establish a process for assigning objects to the category dwarf planet or another status. It also suggested that small solar-system bodies include most solar system asteroids, most TNOs, comets, and other small bodies. In the same meeting, the IAU announced only three members of the dwarf planet category: Ceres, Pluto, and Eris.

Knowledge Gained

The new definitions of planets, dwarf planets, and small solar-system bodies have helped clarify both the nature of these objects and the structure of the solar system, as well as stimulating new research about them. The new definitions have led to new searches for dwarf planets and new research on criteria for hydrostatic equilibrium shape (nearly round) and orbital dominance (clearing the neighborhood).

The IAU maintains a dwarf planet watch-list of about a dozen candidates, which keeps changing as new candidates are found and as more is learned about the physics of existing candidates. Current candidates include the plutinos Orcus and Ixion, cubewanos Quaoar and Varuna, and the SDO Sedna, all of which are similar in size to or larger than Ceres (975 kilometers in diameter) but are not yet established as round. Observations indicate that icy bodies of more than about 400 kilometers reach hydrostatic equilibrium, but rocky objects with more rigid interiors might require at least 800 kilometers. The only other asteroid candidate seems to be Vesta, the second largest at 530 kilometers, which appears to be round except for a large impact crater. The Dawn space probe, scheduled to orbit Vesta by 2011, may resolve its status. Estimates range from forty to two hundred candidates in the Kuiper Belt and more beyond it.

Context

The new definitions of "planet" and "dwarf planet" highlight the increasingly complex nature of the solar system as more is discovered about it. The definitions have also, however, introduced many ambiguities and criticisms.

The new definitions do incorporate accepted theories for the evolution of the solar system and appeal to observational criteria. As planets formed from the dust and planetesimals of the solar disk, their gravity attracted more matter and they eventually dominated their orbits.

However, if planetesimals were sufficiently disturbed by gravitational forces, such as those from nearby Jupiter, they never formed planets and remained as asteroids. Although no planets have completely cleared their orbital neighborhoods, even Mars, as the least dominant planet, has collected more than five thousand times as much material as that which remained in its orbit, while Ceres is only 0.33 times larger and Pluto only 0.07 times larger than the remaining material in their orbits.

Critics complained, however, that the new definitions were arbitrary, since no planet has completely cleared its orbit, and that the round shape of hydrostatic equilibrium is ambiguous, since there are various degrees of roundness. Others voiced concerns about the demotion of Pluto from its longtime status as a planet. Although the National Aeronautics and Space Administration (NASA) decided to accept the new definitions, many respected astronomers, including the director of the New Horizon mission to Pluto, Alan Stern, remained opposed, and his team continued to refer to Pluto as a planet. The discussions and debates would continue at later meetings of the IAU and as more was learned about solar-system objects and their physics.

Joseph L. Spradley

Further Reading

Freedman, Roger A., and William J. Kaufmann III. *Universe*. 8th ed. New York: W. H. Freeman, 2008. College-level introductory text covering the field of astronomy. Contains descriptions of astrophysical questions and their relationships.

Hartmann, William K. *Moons and Planets*. 5th ed. Belmont, Calif.: Thomson Brooks/Cole, 2005. An excellent introductory college text on planetary science by one of the leaders in the field. It has good chapters on the formation of the solar system and on asteroids and other small solar-system bodies.

Serge, Brunier. *Solar System Voyage*. Translated by Storm Dunlop. New York: Cambridge University Press, 2000. This well-illustrated book describes the solar system and discusses issues related to the definition of planets.

Sobel, Dava. *The Planets*. New York: Viking, 2005. A very readable account by a popular science writer of the nature and history of planets and asteroids, and of the scientists who study them.

Soter, Steven. "What Is a Planet?" *Scientific American* 132, no. 6 (January, 2007): 2513-2519. A planetary scientist discusses the controversy over the revised definition of a planet, including both its flaws and the scientific advantages of the concept of a dwarf planet.

Weintraub, David A. *Is Pluto a Planet? A Historical Journey Through the Solar System*. Princeton, N.J.: Princeton University Press, 2006. This book traces the concept of a planet from antiquity to the present day, providing the historical and astronomical context for deciding if Pluto is a planet.

EARTH-MOON RELATIONS

Categories: Earth; Planets and Planetology

The Moon is the closest astronomical body to the Earth, with a mass approximately 1.2 percent that of Earth. This unusually large fraction gives the Moon significant influence over the orbital and rotational motion of Earth, creating tides strong enough to have important geologic and oceanographic effects, among them variations in the length of the day. The Moon, along with the Sun, causes Earth's spin axis to precess with a period of 26,000 years.

Overview

The Moon is the most prominent astronomical body after the Sun. It is the closest astronomical body to the Earth, orbiting at an average center-to-center separation of 384,000 kilometers. The Moon has a radius of 1,740 kilometers; at this distance, it appears to be 0.5° in angular width. The mass of the Moon is 3.74×10^{22} kilograms and the density of the Moon is 3.3 grams per cubic centimeter. Earth, by contrast, has a mass of 5.97×10^{24} kilograms and a radius of 6,380 kilometers, giving it a density of 5.5 grams per cubic centimeter, substantially more than that of the Moon. The lower density of the Moon, along with its lack of a magnetic field, argues that the Moon lacks a molten metallic core such as the Earth has.

Earth is close enough for material thrown off of the Moon by meteorite impact (called "ejecta") to fall onto it. A small number of meteorites discovered in desert areas or in Antarctica closely resemble lunar rocks collected by the Apollo astronauts and have been verified as of lunar origin.

The Moon and Earth are gravitationally bound. They orbit around a common point, called the barycenter, with a period of 27.3 days. This period is called the sidereal month and represents the time for the Earth-Moon system

The Clementine spacecraft captured this image of the Moon in earthshine. The Sun peeks from behind the Moon, with Venus in the far background as the pearl of light near the top. (NASA)

to complete one rotation with respect to the stars. The synodic month, by contrast, is 29.5 days, the time between successive full Moons.

The Earth-Moon system is gravitationally bound to the Sun. Hence, the barycenter orbits the Sun in obedience to Johannes Kepler's three laws of planetary motion: the orbit of the barycenter is an ellipse with the Sun at one focus; the line from the center of the Sun to the barycenter sweeps out equal areas in equal times; and, the cube of the radius of the barycenter orbit is proportional to the square of the period. The barycenter lies on a line joining the center of Earth to the center of the Moon, at a point 4,680 kilometers from the center of Earth. This distance is 73 percent of the radius of Earth. An observer on Mars would see Earth displaced from the ideal elliptical orbit by as much as, of its diameter.

The motion of Earth about the barycenter is superimposed on the elliptical motion of the barycenter about the Sun in a complicated manner. Earth oscillates back and forth across the barycenter ellipse, spending half of a synodic month inside the ellipse (toward the Sun) and the other half outside the ellipse. Simultaneously, the Earth oscillates above and below the ecliptic (the plane of Earth's orbit) spending half of a sidereal month above the plane and the other half below it. These back-and-forth and up-and-down oscillations are not necessarily synchronous. When the Earth is inside the ellipse, the Moon is outside it, and vice versa. A similar arrangement holds for the up-and-down displacements. Absent the Moon, the center of the Earth would coincide with the barycenter and the planetary motion of the Earth would be close to the elliptical ideal. The Earth-Moon system, on the other hand, has one of the most complicated motions in the solar system.

The Earth-Moon system is like an unbalanced dumbbell tumbling end over end about the barycenter. The gravitational pull is the "bar" holding the dumbbell together. The sides of the Earth and Moon facing each other can be referred to as the "inner" sides, and the opposite sides of each can be referred to as the "outer" sides. The gravitational force falls off with distance, making the gravitational pull on the inner side of each body stronger than the gravitational pull on the outer side. This inequality of forces is referred to as gravitational tidal force. Neither the Earth nor the Moon is rigid. Each is plastic enough to change shape under the influence of the tidal force. The gravitational pull of the Moon raises a bulge in the Earth more or less directly under the Moon; the bulge is matched by a similar one at a location more or less directly opposite the Moon. The bulge in the ocean presents itself as the familiar tides. Similar but less familiar tides exist in the atmosphere and in the Earth's crust. The rotation of the Earth attempts to carry these bulges away from the point directly under the Moon, resulting in a slight sideways component to the mutual gravitational pull. This sideways pull acts as a brake on the rotational motion of the Earth, slowing it down and increasing the length of the day. The increase is approximately one-thousandth of a second per century, but it has been accumulating since the creation of the Moon billions of years ago. Growth-ring counts in fossil coral from 400 million years ago seem to indicate

that the year (whose length should not change) consisted of about 400 days back then; now a year consists of about 365 days. In other words, the length of the day has increased by about 10 percent in the past 400 million years.

The sideways pull on the Moon is in the direction of its orbital motion around Earth. Extra energy imparted by the pull increases the radius of the Moon's orbit and also increases the length of the siderial month. Since the length of the day is increasing faster than the length of the sidereal month, eventually the two will become equal and the day and month will be the same. At that time, the Earth will always present the same face to the Moon, just as the Moon always presents the same face to the Earth today.

The rotation of the Earth gives it an oblate shape that is thicker at the equator than through the poles. The gravitational pull of the Sun and Moon on this equatorial bulge acts as a torque that causes the Earth to precess like a top. The spin axis of the Earth currently points toward Polaris, the pole star, but this is only an accident of history. In 13,000 years, Vega (in the constellation Lyra) will be the pole star.

The Galileo spacecraft returned separate images of Earth and its moon that were later compiled into this montage. (NASA/JPL)

Knowledge Gained

The bulk of Earth-Moon interactions are gravitational and are known from earthbound observations. The apparent location of the Sun in the zodiac on the first day of spring (recognized as the day that the Sun rose due East and set due West) held great cultural and religious significance to ancient civilizations and was monitored closely. Over the centuries, it became clear that this location, originally in the constellation Taurus, had moved to the constellation Aries. The Greek astronomer Hipparchus discovered this fact about 130 B.C.E. and from it deduced the 26,700-year circular motion of the north celestial pole. In 1530, Nicolaus Copernicus recognized this as due to drift of the Earth's rotational axis with respect to the fixed stars, and Sir Isaac Newton in 1687 showed the phenomenon to be an effect of Moon's gravitational influence on the Earth.

Edmond Halley in 1693 and Immanuel Kant in 1754 used Newtonian gravitational theory to calculate the locations, dates, and times of total solar eclipses discussed in ancient Greek and Roman documents. Their calculations argued that the eclipses could not have taken place at the dates and places recorded. The discrepancies were eventually traced to changes in the length of the day due to tidal braking.

Starting with Apollo 11, each subsequent lunar landing mission (except the ill-fated Apollo 13) brought back significant amounts of lunar rock for scientific study. Oxygen derived from the lunar material proved to have the same ratio of isotopes as oxygen found on Earth. In contrast, oxygen retrieved from meteorites believed to be of Martian origin had substantially different isotopic ratios.

This discovery, in conjunction with the observation that the Moon lacks an iron core, led to the impact theory of lunar origin. In this theory, the young Earth and a body approximately the size of Mars collided some 4.5 billion years ago. The collision threw a substantial amount of the Earth's crust into space, where some of the material coalesced into the Moon, with the remainder falling back to Earth. Since this happened after the bulk of the iron in the proto-Earth had sunk into the core, the material that formed the Moon was relatively iron-free.

Context

The combined motion of the Earth and the Moon around their common barycenter is one of the most complicated problems in celestial mechanics. Newton once referred to it as the only problem that ever gave him a headache.

Several factors complicate the solution. The influence of the Sun makes the problem a three-body gravitational interaction rather than the simpler two-body problem conquered by Kepler. Unlike the two-body problem, the three-body problem cannot be solved in closed analytic form; particular approximate solutions exist for special configurations, but the Sun-Earth-Moon trio does not conform to any of them.

The Earth and Moon are also too close for either to be regarded as point masses. Further, neither is purely spherical: the Earth is ellipsoidal, with an equatorial bulge as a product of its rotation; the Moon is oval as a result of a permanent tidal bulge on the side facing the Earth. The rotational period of the Moon equals its orbital period, so that one face perpetually faces the Earth, but the orbit is not circular, so that the Moon moves along the orbit at a varying rate. This causes the side of the Moon facing the

The Moon viewed from Earth orbit, with the Earth's atmosphere rising above the planet's surface in the foreground. (NASA)

The Earth viewed from the Moon, as the Apollo 8 astronauts began their orbit on December 29, 1968. (NASA)

Earth to rock back and forth, a motion known as libration. The deviation from circularity (called the eccentricity) is itself variable, driven by the gravitational pull of the Sun, so that the the extent of the libration waxes and wanes. This variation in eccentricity is called evection.

Billy R. Smith, Jr.

Further Reading

Brusche, P., and J. Sundermann. *Tidal Friction and the Earth's Rotation*. Berlin: Springer, 1978. A scholarly work. The first paper, "Pre-Telescopic Astronomical Observations" by F. R. Stephenson covers the historical eclipse data that revealed the slow increase in the length of the day due to tidal braking.

Comins, N. *What if the Moon Didn't Exist? Voyages to Earths That Might Have Been*. New York: HarperCollins, 1993. An astronomer examines how the Earth might have evolved without the interaction of a massive nearby Moon. An unusual but entertaining and engaging exploration of Earth-Moon interactions. Index and bibliography.

Darwin, G. *The Tides and Kindred Phenomena in the Solar System*. San Francisco: W. H. Freeman, 1962. A thorough discussion of tides from all perspectives: oceanographic, hydrodynamic, geological, astronomical, and historical. Contains a chapter on the less familiar tides in the atmosphere and in the Earth's crust. The remarks on tidal coupling between the Sun and planets are unfortunately out of date.

Ferguson, Kitty. *Tycho and Kepler: The Unlikely Partnership That Forever Changed Our Understanding of the Heavens*. New York: Walker, 2002. This book engagingly describes what is probably the most fruitful and important collaboration in all of the physical sciences. Tycho Brahe's naked-eye observations were the most accurate ever obtained before the invention of the telescope. They revealed hitherto unknown variations in the motion of the Moon; Brahe's observations of Mars made it possible for Kepler to discover the laws of planetary motion that bear his name. Well illustrated (includes six pages of color plates), with notes, bibliography, and index.

Kolerstrom, Nicholas. *Newton's Forgotten Lunar Theory: His Contribution to the Quest for Longitude*. Santa Fe, N.Mex.: Green Lion Press, 2000. Before the invention of the ship's chronometer, navigators used the motion of the Moon to determine longitude. Doing so accurately requires a very accurate theory of the Moon's orbital motion. In Newton's time, this was an area of immensely important scientific research. This book outlines his efforts at solving this exceedingly difficult problem. Chapter 1 describes the motions of the Sun, Earth, and Moon.

Moore, Patrick. *On the Moon*. London: Cassell, 2001. Contains an excellent nontechnical discussion of lunar motion and tides.

ECLIPSES

Categories: Earth; The Solar System as a Whole

Eclipses, occultations, and transits occur when three celestial bodies line up, causing the middle body to block the path of light between those on the two ends. In particular, solar and lunar eclipses witnessed from Earth are spectacular phenomena that have been objects of awe, study, and speculation since ancient times. Once they were understood, they became powerful tools of science used to investigate topics as diverse as geodesy and general relativity.

Overview

Eclipses of the Sun and Moon are impressive events. They have captivated people since before recorded history and continue to excite us today. Once considered great omens or portents, they have become among the most powerful means with which science tests theories in a remarkable variety of areas.

A lunar eclipse (or eclipse of the Moon) occurs when the Moon passes through the shadow of the Earth. For this to happen, the Moon must be on the side of the Earth opposite the Sun, at the time when the Moon's phase is full.

A solar eclipse (or eclipse of the Sun) occurs when the Moon passes between the Earth and the Sun, blocking all or part of the Sun as seen from Earth. For this to happen, the Moon's phase must be new.

Not every full or new Moon results in an eclipse, since the orbit of the Moon lies in a plane tilted about 5° to the ecliptic plane, the plane containing Earth's orbit around the Sun and in which the Earth and the Sun always lie. Unless a full or new Moon occurs when the Moon is very close to crossing the ecliptic plane, the Earth's shadow will miss the Moon (at full Moon) or the Moon's shadow will miss the Earth (at new Moon), and no eclipse will take place. This condition has been known and used to predict eclipses since ancient times and is the source of the name of the ecliptic plane.

A lunar eclipse is visible from all points on Earth where the Moon is above the horizon. It may be either umbral or penumbral. During an umbral lunar eclipse, at least part of the Moon passes through the Earth's umbra, the dark inner shadow in which the Earth blocks light coming from all parts of the Sun. If the entire Moon passes through the umbra, it is called a total lunar eclipse; if only part of the Moon passes through the umbra, it is called a partial lunar eclipse. During a penumbral lunar eclipse, the entire Moon misses the umbra and passes only through the Earth's penumbra, a region of partial shadow surrounding the umbra in which Earth cuts off light from some but not

A partial solar eclipse was visible from Earth as the Moon obscured the Sun in 1994. (© Sébastien Gauthier/NASA)

all parts of the Sun. An observer on the Moon during a penumbral lunar eclipse would see part of the Sun covered by the Earth and part of the Sun extending beyond the edge of the Earth.

The Moon is dimmed slightly while in the penumbra, but it does not darken appreciably unless it enters the umbra. When the Moon enters the umbra, the previously bright surface of the full Moon darkens to a much dimmer reddish glow, illuminated only by sunlight that has been refracted and scattered around the Earth by Earth's atmosphere. The brightness and color of this illumination can vary markedly from one umbral lunar eclipse to another (from orangish red to a dull reddish brown to a ghostly brownish gray), depending on the atmospheric conditions on Earth. Occasionally, some areas of the Moon will seem less illuminated than others.

A total lunar eclipse can last several hours. From the time the Moon begins to enter the umbra, it takes about an hour for the eclipse to become total. Totality can last for nearly two hours, with another hour required for the Moon to leave the umbra entirely. The limb (or edge) of the Moon nearest the observer's eastern horizon enters the Earth's shadow first, and at the end of the eclipse it is this limb that brightens first.

Solar eclipses are somewhat more complex. By coincidence, the Sun and the Moon both have nearly the same apparent size or angular diameter—about one-half degree of arc—as viewed from Earth. The Sun's actual diameter is about 400 times larger than the Moon's, but the Sun is also about 400 times farther away from the Earth than the Moon is. Because the orbit of the Moon around the Earth and the orbit of the Earth-Moon system around the Sun are both slightly elliptical, the apparent (angular) size of the Moon and Sun varies as seen from the Earth. On average, the angular diameter of the Moon, 0.518°, is slightly less than the angular diameter of the Sun, 0.533°, as seen from Earth. However, when the Moon is at perigee (its closest approach to Earth), its angular diameter increases to 0.548°. When the Earth is at perihelion (its closest approach to the Sun), the Sun's angular diameter increases to 0.542°. Thus, when the Moon is near perigee, its angular size always exceeds that of the Sun, and if a solar eclipse occurs then, the Moon can completely cover the Sun.

The Moon's umbra is conical in shape, with its base at the Moon and narrowing to its apex (or tip) as it nears Earth. If the Moon is near perigee, the apex of its umbra will fall inside the Earth, and observers in the region of the Earth's surface within the umbra will see the silhouette

of the Moon completely cover the Sun's visible surface (its photosphere); this is called a total solar eclipse. The region of the Earth's surface within the umbra at any moment is quite small, never more than a few hundred kilometers across. Due to the orbital motion of the Moon around the Earth and the Earth's rotation on its axis, its umbra sweeps a path (the eclipse track) thousands of miles long across the Earth's surface from west to east at speeds always exceeding 1,700 kilometers per hour relative to the Earth's surface. The maximum duration possible for totality is about seven and a half minutes, although the complete eclipse, including the partial phases at the beginning and end as the Moon slowly covers and then uncovers the Sun, may take more than four hours.

Because the Moon's average angular size is slightly smaller than the Sun's average angular size, the tip of the Moon's umbra will not always reach the Earth's surface during a solar eclipse. If this is the case, the Moon will not completely cover the Sun's photosphere. When the Moon is centered on the Sun, a narrow ring, or annulus, of the Sun's bright photosphere remains visible around the Moon's silhouette. This is called an "annular solar eclipse." Annular solar eclipses are about 20 percent more frequent than total solar eclipses.

Beyond the region of the Earth's surface in which a total or annular solar eclipse is seen, there is an area thousands of kilometers wide inside the Moon's penumbra. Within this area, the silhouette of the Moon covers part but not all of the Sun's photosphere. This is called a partial solar eclipse.

Occultations and transits are phenomena similar to eclipses in which the apparent angular size of the body in front is substantially larger or smaller than the apparent angular size of the body in back. An apparently large body moving in front of an apparently smaller one is called an occultation. The Moon frequently occults bright stars, which are seen to wink out instantly when they pass behind the limb of the Moon. The Moon also occasionally occults planets, and planets are seen to occult their satellites and stars. (Actually, other planets can both occult and eclipse their satellites as seen from Earth. The distinction between the two phenomena is that the satellite moving behind the planet is an occultation, while the satellite passing through the planet's shadow is an eclipse. The two events are not necessarily precisely coincident in time, because the planet's shadow cone will not be directly behind the planet as seen from Earth unless the planet is on the side of the Earth almost exactly opposite the Sun.) The Sun also occults objects, but because

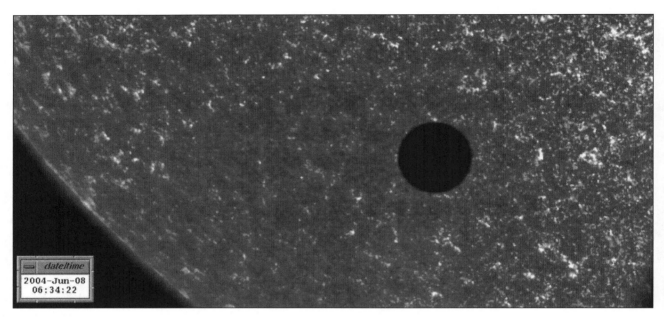

The planet Venus can be seen crossing the solar face during its 2004 transit. (NASA)

of the Sun's brightness, such solar occultations cannot be seen at visible wavelengths; however, they have been observed at radio wavelengths when the object being occulted is a source of radio emission. An apparently small body moving in front of an apparently larger one is called a transit. On rare occasions it is possible to witness from Earth the transit of the planets Venus or Mercury across the Sun. When this happens, the planet appears like a small, black dot moving across the face of the Sun. Transits of satellites and their shadows across their parent planets also can be observed.

Another, related phenomenon is that of eclipsing binary stars. Some of the stars seen in the sky are in fact pairs of stars orbiting about one another. If the Earth lies near the plane of their mutual orbit, the combined light of the system is seen to vary in brightness as the stars alternately block each other from Earth's view. By observing these variations in brightness, astronomers can determine some of the characteristics of the individual stars (such as relative sizes and surface temperatures) and study their interactions.

Methods of Study

Ancient peoples, who used astronomical observations to keep track of planting seasons and the like, usually imputed magical or spiritual significance to eclipses and consequently tried to predict them. They did this by watching the changing position of the Moon against the background of the stars and by recording patterns in the recurrence of eclipses.

A lunar, or synodic, month is the period from one new Moon to the next, about 29.53 days. The draconic month is the time required for the Moon to complete one cycle of crossing and recrossing the ecliptic plane (from south to north and from north to south), about 27.21 days. The coincidence of these two cycles produces eclipses, so the pattern of eclipses starts again when the two cycles return to the same relative matchup. This happens every 223 lunar months (equaling 6,585.32 days or 242 draconic months or a little more than 18 years), in a repeating pattern called the saros cycle. The cycle lasts 18 years and 11.32 or 10.32 days, depending on whether 4 or 5 leap years occur during that time. The extra third of a day (or about 8 hours) means that eclipses 223 lunar months apart, although similar in overall geometry, will occur about 120° farther west in longitude because of the Earth's rotation. After three such saros cycles, about 56 years and 1 month, eclipses repeat in nearly the same part of the Earth again.

These cycles, at least as they applied to lunar eclipses, were known to Babylonian astronomers by around the eighth century B.C.E. and may have been known to some peoples long before that (based on disputed interpretations of a circle of fifty-six pits around the neolithic monument at Stonehenge in England). Knowledge of these cycles enabled the Babylonians to predict the relative motions of the Sun and Moon in the sky. The saros cycle was of limited use in predicting solar eclipses, because the path of totality is so narrow. Very precise knowledge of the relative motions of the bodies involved was required. This was not possible before Sir Isaac Newton developed his laws of motion and gravity along with the calculus in the seventeenth century. One of the first tests he applied to his new methods was the calculation of the orbit of the Moon.

Centuries of refinement, both of mathematical methods and of measurements of the positions of the Moon, Earth, and Sun relative to one another, were necessary to achieve modern accuracy in eclipse predictions. Astronomers can now calculate eclipses, including exact times and paths of totality, many years into the future with almost total precision. However, even these calculations are limited by residual uncertainties in the motions of the bodies involved when extrapolations hundreds of years in the past or future are attempted. With three bodies gravitationally interacting, no exact solution for the orbits is possible, although modern approximation methods are very good. Furthermore, the rate of rotation of the Earth has varied over time and continues gradually to slow, complicating the calculation of eclipse times and locations far back into the past or far forward into the future.

A total solar eclipse is perhaps the most spectacular natural event that can be seen. During a total solar eclipse, the Moon appears as a dark disk that slowly moves across and covers the bright disk of the Sun. Just before the Sun is completely covered, the remaining bright crescent narrows until it becomes a chain of bright spots along the edge of the Moon. These spots, called Baily's beads, represent a last glimpse of the Sun's photosphere between mountains at the edge of the Moon. Then for a few seconds, the Sun's chromosphere (a thin layer of transparent gas above the photosphere) can be seen as a red fringe along the leading edge of the Moon. At about this time, rapidly moving shadow bands, striations of light and dark a

few centimeters across, can be seen rippling across the ground and along walls. These are believed to be due to atmospheric refraction. The sky turns dark during totality, but not completely dark; some light is scattered into the umbra from outside the region of totality. The darkness at totality produces uneasiness among some animals, and birds are sometimes seen to go to roost, as at sunset. The solar corona, the Sun's outer atmosphere of hot ionized gases, is seen as a glowing white halo around the dark silhouette of the Moon. Smaller, fiery red solar prominences are often observed around the edge of the Moon. As totality ends, the shadow bands, chromosphere, and Baily's beads can be seen briefly again.

It is important never to look directly or through optical instruments such as telescopes, binoculars, or camera viewfinders at the uneclipsed or partially eclipsed Sun without using suitable filters manufactured for this purpose. Common sunglasses are insufficient to prevent severe, painful, and permanent eye injuries, which can occur nearly instantaneously. However, during totality,

The Hubble Space Telescope captures three of Jupiter's moons—Io, Ganymede, and Callisto—crossing the gas giant's face and casting shadows on its surface in a rare triple eclipse. (NASA/ESA/E.Karkoschka, University of Arizona)

when the Moon completely covers the Sun's photosphere and only the corona is visible, no filter is needed. Furthermore, no filters are needed to watch lunar eclipses. A telescope or at least binoculars generally are needed to enlarge the view of most occultations and transits.

Context

Eclipses, occultations, and transits have been powerful tools to derive useful information, to make new discoveries, and to test and confirm various theories and predictions. One of the oldest such uses can be traced back to the ancient Greeks. They understood how lunar and solar eclipses occurred, and the circular outline of the Earth's shadow seen on the Moon during lunar eclipses was cited by Aristotle (384-322 B.C.E.) and other Greek philosophers as evidence that the Earth must be spherical.

In 1675, the Danish astronomer Ole Rømer was studying the orbital motion of Jupiter's four largest moons (its Galilean satellites) by carefully timing their eclipses in Jupiter's shadow. He discovered that the eclipses occurred later than expected when Earth and Jupiter were farther apart and earlier than expected when Earth and Jupiter were closer. He realized that this phenomenon could be explained if light did not travel instantaneously but took time (about 16.6 minutes by his measurements) to cross the Earth's orbit (a distance of 2 astronomical units). Since the length of the astronomical unit was not known accurately then, he never actually calculated the speed of light in "everyday" type units, but this was the first demonstration that light traveled at a finite speed.

Eclipses of the satellites of Jupiter, as well as much less frequent lunar and solar eclipses, helped seafarers find their location at sea and map the Earth. Although latitudes can be determined easily by measurements of the maximum altitude above the horizon reached by the Sun in the daytime or specific stars at night, longitudes cannot be determined astronomically without a time reference. Pendulum clocks did not run accurately at sea because of the motion of the ship, but the calculated times of eclipses provided the necessary time reference and permitted reliable navigation and the construction of accurate maps.

Solar eclipses have helped resolve some of the most important questions in science. They provided an infrequent opportunity for observing some of the Sun's features—such as its corona, chromosphere, and prominences—which, before the development of modern instruments, were otherwise hidden most of the time by the brightness of the Sun. The element helium (from Helios,

the Greek Sun god) was first discovered in the flash spectrum of the chromosphere during a total solar eclipse in 1868.

An observational test of general relativity was carried out first during the 1919 total solar eclipse. In 1915, Albert Einstein had published his then-controversial general theory of relativity. One of its predictions was that light passing by a massive object, such as the Sun, would be deflected by gravity. By photographing the star field around the totally eclipsed Sun and comparing the apparent positions of the stars near the edge of the Sun to a photograph of the same star field when the Sun was in a different part of the sky, astronomers verified this prediction of general relativity. It has been confirmed repeatedly at several total solar eclipses since then. Radio interferometry observations of a shift in the position of quasar 3C273 when it is occulted by the Sun have provided even more accurate confirmation.

In 1977, astronomers observed the occultation of a star by the planet Uranus to study its atmosphere by the way it absorbed light from the star. They were surprised when the star faded and brightened several times before and after being occulted by Uranus itself. This was attributed to a set of rings around Uranus, and was the first discovery of rings around a planet other than Saturn.

Firman D. King,
revised by Richard R. Erickson

Further Reading

Baker, Robert H. *Astronomy*. 7th ed. Princeton, N.J.: Van Nostrand, 1959. Baker's classic astronomy text provides a complete and lucid description of eclipse phenomena, along with related issues in spherical astronomy.

Brewer, Bryan. *Eclipse*. 2d ed. Seattle: Earth View, 1991. A good introduction to eclipse phenomena and their history. For the general reader.

Chaisson, Eric, and Steve McMillan. *Astronomy Today*. 6th ed. New York: Addison-Wesley, 2008. A well-written college-level textbook for introductory astronomy courses. Has several pages on eclipses.

Fraknoi, Andrew, David Morrison, and Sidney Wolff. *Voyages to the Stars and Galaxies*. Belmont, Calif.: Brooks/Cole-Thomson Learning, 2006. A well-written, thorough college textbook for introductory astronomy courses. Includes a section on eclipses.

Freedman, Roger A., and William J. Kaufmann III. *Universe*. 8th ed. New York: W. H. Freeman, 2008. College-level introductory astronomy textbook, thorough and well written. A major part of one chapter deals with eclipses.

Schneider, Stephen E., and Thomas T. Arny. *Pathways to Astronomy*. 2d ed. New York: McGraw-Hill, 2008. A thorough college textbook for introductory astronomy courses, divided into many short sections on specific topics. Eclipses are discussed in several sections.

Stevenson, F. Richard. "Historical Eclipses." *Scientific American* 247 (October, 1982): 170-183. This article discusses how historical records and astronomical calculations are compared to resolve questions in both history and astronomy.

Taff, Laurence G. *Celestial Mechanics*. New York: Wiley-Interscience, 1985. This book describes how orbits are calculated and makes it clear why the process becomes so complicated when long time intervals are involved.

Zirker, Jack B. *Total Eclipses of the Sun*. Expanded ed. Princeton, N.J.: Princeton University Press, 1995. Zirker explains how solar eclipses are observed and describes some of the scientific results that have been obtained from them.

ERIS AND DYSNOMIA

Category: Small Bodies

The scattered disk object Eris is the largest dwarf planet and the most distant solar-system object astronomers have identified. Observation of Eris so far has revealed only one satellite, Dysnomia. Eris was initially hailed as a possible tenth planet when discovered in 2005, but debate within the astronomical community regarding what constitutes a planet led the International Astronomical Union to categorize it as a dwarf planet.

Overview

Eris, formally designated (136199) Eris, is the largest of the known trans-Neptunian objects. As of 2009, only one satellite, Dysnomia, had been found in orbit around Eris. The most distant solar-system object observed to date, Eris is classified as a scattered disk object. Despite its remoteness, it can be viewed with powerful amateur equipment. At aphelion, Eris lies beyond the outermost region of the Kuiper Belt; at perihelion, it passes within the range of Neptune's influence. As of 2009, Eris was one of only three known plutoids (the others are Pluto

and Makemake) and four known dwarf planets (the three plutoids plus Ceres). Eris is larger than Pluto, its orbital period is more than twice Pluto's, and at aphelion it is roughly three times farther from the Sun than Pluto.

Researchers have calculated Eris's orbit using archival data collected before the dwarf planet's discovery. Once Eris was identified in 2005, the bright, slow-moving object was easy to spot in images going back to 1989. Eris takes 557 years to travel around the Sun, following a highly elliptical path with a semimajor axis, or mean orbital radius, of 67.9 astronomical units (AU), or more than 10 billion kilometers. When Eris was discovered, it was near its aphelion of 97.5 AU (more than 14.5 billion kilometers) from the Sun. It will reach perihelion at 38.2 AU (more than 5.7 billion kilometers) in the mid-twenty-third century. For the most part, the orbit of Eris is typical of a scattered disk object. However, Eris's orbital inclination is an atypically high 44° from the solar system's orbital plane. The inclination and period of Eris's rotation have yet to be ascertained, although it is currently thought that a day on Eris is about eight hours long.

Images from the Hubble Space Telescope indicate that Eris has a diameter of roughly 2,400 kilometers, making it slightly larger than Pluto. Spitzer Space Telescope observations suggest that Eris's diameter may be as great as 2,600 kilometers. Calculations based on Keck Observatory and Hubble Space Telescope observations of Dysnomia's orbit around the dwarf planet indicate that Eris has a mass of 1.66×10^{22} kilograms, or 3.66×10^{22} pounds (27 percent greater than Pluto's) and a bulk density of 2.3 grams per cubic centimeter (0.83 pound per cubic inch).

Using Eris's density, researchers have determined that the dwarf planet's interior is most likely similar to Pluto's—that is, about half rock and half ice. Where Pluto's surface appears reddish and partly rocky, however, Eris has a uniform, highly reflective, almost gray surface. Near-infrared images obtained at the Gemini North Observatory in Hawaii indicate that this surface is predominantly frozen methane. In Pluto's case, darker surface hues are attributed to tholins, reddish-brown breakdown products formed when methane and similar organic compounds are subjected to solar ultraviolet irradiation. Where tholin deposits make Pluto's surface darker, the albedo is lower and the temperature higher. Methane ice melts away from these comparatively warm patches. By contrast, methane ice appears to envelop Eris in a bright, near-uniform coating. This suggests not only that Eris remains cold enough that

An artist's view of dwarf planet Eris and its satellite Dysnomia, based on data from the Hubble Space Telescope. (NASA/ESA/Adolph Schaller for STScI)

its methane stays in a frozen state but also that there may be a subsurface source of methane that replenishes Eris's surface coating of methane ice and covers up whatever tholins are deposited.

As Eris moves from aphelion to perihelion, its temperature increases from its current value of 30 kelvins (-243° Celsius) to 56 kelvins (-217° Celsius). It is possible that, when Eris approaches the Sun, some of its surface ices become warm enough to sublimate and form a thin atmosphere. Whatever gases do not escape the atmosphere freeze once again as the dwarf planet moves toward aphelion.

To date, only one satellite has been observed in orbit around Eris: Dysnomia, known technically as (136199) Eris I. Hubble Space Telescope and Keck Observatory data indicate that Dysnomia is roughly 150 kilometers in diameter. It takes about 16 days for Dysnomia to complete its near-circular orbit around Eris at a distance of approximately 37,000 kilometers. The small satellite is believed to be composed largely of frozen water.

Astronomers Mike E. Brown of the California Institute of Technology, Chad A. Trujillo of Gemini Observatory, and David Rabinowitz of Yale University discovered Eris through an ongoing survey conducted at Palomar Observatory in southern California using the Samuel Oschin telescope. Images taken on the night of October 21, 2003, showed the large, bright object traveling slowly across the sky. Its movement was slow enough, in fact, that Eris went undetected when the images were first analyzed. The researchers' discovery in November, 2003, of Sedna, another large and slow-moving trans-Neptunian object, led them to adjust their detection scheme and reanalyze their survey data. On January 5, 2005, they identified the planet-sized scattered disk object that would later be known as Eris, designating it 2003 UB313.

Brown and his team intended to follow standard scientific protocols by verifying their discovery, studying it, documenting it thoroughly, and making it known through a scientific paper published in a reputable journal. However, in July, 2005, they learned that detailed records of their telescope use had inadvertently been made accessible to anyone with Internet access, and that an abstract they had recently published unwittingly contained clues about where in the sky to look for their recent—and unannounced—trans-Neptunian discoveries. When, five days after the abstract was issued, researchers in Spain announced the discovery of 2003 E161—a trans-Neptunian object Brown and his colleagues had also found—it appeared the team had to lay claim to 2003 UB313 or risk having someone else take credit for finding it.

The California Institute of Technology, the Jet Propulsion Laboratory, and the National Aeronautics and Space Administration (NASA) announced the discovery of 2003 UB313 on July 29, 2005, in press releases that referred to the object as the tenth planet. The media was quick to adopt the team's nickname for the newly discovered member of the solar system: Xena, so called for the heroine of the television series *Xena: Warrior Princess*. (Some overenthusiastic journalists, seeing "planetlila" in Brown's Web address, pounced upon Lila as the new planet's name, only to learn that the URL was a whimsical tribute to Brown's newborn daughter.)

That same year, Brown, Trujillo, and Rabinowitz collaborated with the engineering team at the Keck Observatory on Mauna Kea, Hawaii, to search 2003 UB313 and three more of the brightest trans-Neptunian objects for satellites. Using the Keck's new Laser Guide Star Adaptive Optics system, which enabled the researchers to view details as precise as those seen from the Hubble Space Telescope, they found S/2005 (2003 UB313) 1—the faint satellite that would come to be called Dysnomia—on September 10, 2005. The team dubbed the satellite Gabrielle, after Xena's television-show sidekick.

In the summer of 2006, at a meeting of the General Assembly of the International Astronomical Union (IAU), criteria were developed regarding what constitutes a planet, and a new category of "dwarf planet" was established. Under the IAU's new definitions, 2003 UB313 was not a planet but rather a dwarf planet. On September 6, 2006, the discovery team proposed to the IAU that

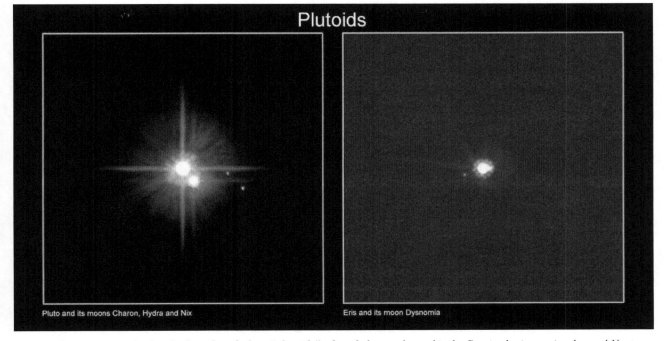

Plutoids

Pluto and its moons Charon, Hydra and Nix

Eris and its moon Dysnomia

Pluto and Eris, among the first bodies classified as "plutoids": dwarf planets that orbit the Sun in the icy region beyond Neptune. (International Astronomical Union)

dwarf planet 2003 UB313 be named Eris and its moon be named Dysnomia. The IAU accepted and announced the names one week later on September 13, 2006.

Knowledge Gained

The 2005 discovery of 2003 UB313 added fuel to a long-standing and heated debate in the astronomical community over Pluto's status as a planet. The newly discovered object had a greater diameter and was more massive than its distant neighbor Pluto. If Pluto's size and mass were sufficient to qualify it for "planethood," then the new object should likewise be classified as a planet.

The controversy came to a head in August, 2006, at IAU's General Assembly in Prague. At the unusually contentious meeting, members reached the non-unanimous conclusion that a celestial body should be considered a planet if it (1) orbits the Sun (satellites are not planets); (2) is massive enough for self-gravity to shape it into a sphere; and (3) has accreted or scattered other bodies in its neighborhood to clear its orbit. Objects meeting the first two criteria but not the third would be classified as dwarf planets.

Both Eris and Pluto orbit the Sun, and both are spherical, yet neither has cleared the vicinity around its orbit. With the IAU's official acceptance of the new definitions on August 24, 2006, 2003 UB313 ceased to be a possible tenth planet and instead became the largest known dwarf planet. Likewise, Pluto was "demoted" to dwarf planet status after more than seven decades of being regarded as a planet. Ceres was determined to be a dwarf planet that might also be an asteroid.

2003 UB313 received its official name, Eris, on September 13, 2006. In Greek mythology, Eris was the goddess of discord and strife. Given the uproar the object's discovery caused among astronomers—and a public accustomed to a nine-planet solar system—it was aptly named. The dwarf planet's satellite was named after the mythological figure Dysnomia, the goddess Eris's daughter and the spirit of lawlessness. The name Dysnomia was also a tip of the hat to Eris's old nickname—the television actress who played warrior princess Xena was Lucy Lawless.

Almost two years later, Eris was assigned to a new subcategory of dwarf planet. On June 11, 2008, the IAU announced that dwarf planets orbiting the Sun at a semimajor axis greater than that of Neptune would be known as "plutoids." Eris and Pluto qualified; Ceres, located in the asteroid belt, did not. In July, 2008, the Kuiper Belt object known as 2005 FY9 received its official name, Makemake, and became the third largest dwarf planet and plutoid.

Context

Discovery in the early 1990's of small objects beyond Neptune inspired researchers such as Brown to look to the Kuiper Belt and beyond for larger celestial bodies. Brown and Trujillo made their first major trans-Neptunian find in 2002 with Quaoar. In 2003, they discovered a larger object, Sedna, then the remotest solar-system object that had ever been found. Sedna's slower motion across the heavens led the team to look for objects moving at even lower rates, which revealed Eris.

Theoretical models suggest that Eris and other comparatively large trans-Neptunian objects in high-inclination orbits were originally near the inner edge of the Kuiper Belt. When they were subsequently scattered into the outer belt and beyond, they achieved orbits with higher inclinations than objects originating in the outer belt. More massive objects that had their origins in the inner belt may now occupy remote, high-inclination orbits. Researchers are now looking for as-yet-undiscovered large objects orbiting at these high inclinations.

At present, there is no space mission planned to Eris. While the uncrewed New Horizons will conduct a flyby of Pluto, the spacecraft's limited maneuvering capability keeps it from passing close enough to observe Eris.

Karen N. Kähler

Further Reading

Brown, M. E., C. A. Trujillo, and D. L. Rabinowitz. "Discovery of a Planetary-Sized Object in the Scattered Kuiper Belt." *The Astrophysical Journal* 635, no. 1 (2005): L97-L100. The journal article in which Eris's discovery team first presented their findings. Explains how the team discovered 2003 UB313 and summarizes the initial physical and orbital characterizations of the object. Includes tables, figures, and references. Technical but straightforward.

_____. "Satellites of the Largest Kuiper Belt Objects." *The Astrophysical Journal* 639, no. 1 (2006): 143-146. The scientific paper in which the discovery team introduced Eris's moon. Describes the survey of 2003 UB313 and three other trans-Neptunian objects for possible satellites. Includes tables, figures, and references. Challenging for the lay reader.

Chang, Kenneth. "Dwarf Planet, Cause of Strife, Gains 'the Perfect Name.'" *The New York Times*, September 15, 2006, p. A20. A brief but informative article on how dwarf planet 2003 UB313 and its satellite came to be named Eris and Dysnomia—and nicknamed Xena and Gabrielle.

_____. "Ten Planets? Why Not Eleven?" *The New York Times*, August 23, 2005, p. F1. Reprinted in *The Best American Science Writing 2006*, edited by Atul Gawande. New York: HarperPerennial, 2006. An engaging profile of team leader Mike Brown, published shortly after the announcement that 2003 UB313 had been found. Includes a chart summarizing five of Brown's major finds, among them 2003 UB313.

Faure, Gunter, and Teresa M. Mensing. *Introduction to Planetary Science: The Geological Perspective*. New York: Springer, 2007. Designed for college students majoring in Earth sciences, this textbook provides an application of general principles and subject material to bodies throughout the solar system. Excellent for learning comparative planetology.

Freedman, Roger A., and William J. Kaufmann III. *Universe*. 8th ed. New York: W. H. Freeman, 2008. College-level introductory text covering the field of astronomy. Contains descriptions of astrophysical questions and their relationships.

Fussman, Cal. "The Man Who Finds Planets." *Discover*, May, 2006, 38-45. A highly accessible account of the discovery of 2003 UB313 (then still popularly known as Xena) and other trans-Neptunian objects, as told by team leader Mike Brown. Includes photographs.

IMPACT CRATERING

Category: Planets and Planetology

Space-age discoveries about the surface character of other terrestrial planets and satellites around planets throughout the solar system reveal that the early Earth must have been heavily scarred by impacts with planetesimals and minor bodies. Erosion processes and plate tectonics have obliterated most of these ancient craters, but new evidence that major impacts may have had a significant role in shaping the evolution of life has spurred a search for large impact craters.

Overview

Impact cratering is one of the most fundamental geologic processes in the solar system. Craters have been found on the surfaces of all the solid planets and natural satellites thus far investigated by spacecraft. Mercury and the Moon, bodies whose ancient surfaces have not been reworked by subsequent geologic processes, preserve a vivid record of the role that impact cratering has played in the past. It is inconceivable that the Earth somehow escaped the bombardment that caused such widespread scarring or that it does not continue to be a target for planetesimals still roaming the solar system.

As recently as the 1960's, only a handful of sites on Earth were accepted to be of impact origin. In the early 2000's, the number of confirmed astroblemes (circular surface features considered to have been large impact craters) was well in excess of one hundred and increasing at the rate of several per year. In addition, many "probable" and "possible" impact features are under study. Nevertheless, an enormous discrepancy exists between the number of identified or suspected impact sites on Earth and the number that might be expected.

It is assumed that the flux of incoming bodies is the same for Earth as it is for the Moon. Making allowances for the fact that the Earth is the largest "target" of any of the terrestrial planets and that more than two-thirds of its surface is covered by water, planetologists calculate that land areas of the Earth should have been scarred by at least fifteen hundred craters 10 kilometers or more in diameter. In actuality, only about half of the known astroblemes are in this size range. On a global scale, 99 percent of the predicted large impact craters seem to be missing. However, this statistic is not a valid indicator of the impact history of the Earth. Although the impact phenomenon is a geographic process, the probability for discovering impact sites is strongly modified by the geologic stability of various regions of the Earth and by the intensity of the search programs in those areas. Roughly one-half of all the confirmed astroblemes have been found in Canada, which constitutes only one percent of the Earth's surface. In part, this is owing to the stability of Precambrian rock of the Canadian Shield, which thus preserves more of the crater's features, but it also reflects a diligent research effort by Canada's Department of Energy, Mines, and Resources. In general, the number of large impact sites found in the well-explored areas of the Earth agrees with the accepted rate of crater formation on the other terrestrial planets in the past two billion years.

The obvious difference between the surface appearances of Earth and the Moon is explained not by any difference in the rate at which impact craters have formed but in the rate at which they are destroyed. Most of the tremendous numbers of craters on the Moon are more than 3.9 billion years old, while the Earth's oldest surviving astroblemes were formed less than 2 billion

A view of the Moon's Mare Imbrium, taken by the Apollo 17 crew in 1972, shows the cratered lunar surface. (NASA)

years ago. Studies have shown that erosion effectively removes all traces of a 100-meter (diameter) crater in only a few thousand years, and that a 1-kilometer-wide crater, such as the well-known Barringer meteor crater in Arizona, will disappear within a million years. Only craters with diameters greater than 100 kilometers can be expected to leave any trace after a billion years of erosion. This explains not only the absence of widespread cratering on Earth's landscape but also the fact that, among the astroblemes known to exist, medium and large scars are more common than small ones.

Significant craters can be produced only by objects having masses of hundreds of thousands to billions of tons. The Barringer crater, 1.2 kilometers wide and 200 meters deep, is believed to have been formed by a one-million-ton planetesimal that was perhaps 50 meters in diameter. A 27-kilometer-wide astrobleme known as the Nördlinger Ries crater in Germany required an impacting body greater than 1 kilometer in diameter with a mass in excess of 1 billion tons. Planetesimals as large as these two examples are not characteristic of the vagrant meteors that wander through the solar system and occasionally streak into the Earth's skies as shooting stars.

Most of the past impacts on Earth and the Moon appear to be attributable to a family of asteroids known as the Apollo-Amor group (after two specific members of the family). Members of this group are in orbits that graze Earth's orbit and become subject to orbital perturbations that lead them across Earth's path periodically. It is estimated that the average Apollo-Amor object intersects Earth's orbit once every five thousand years, although usually the planet is at some other point on its orbit when this happens. The probability of a collision between Earth and any given Apollo-Amor object is small, but several studies have shown that this family contains between 750 and 1,000 asteroids larger than 1 kilometer in diameter. Statistical analysis suggests that such sizable bodies must collide with the Earth an average of once every 600,000 years.

Impact events involve tremendous transfers of energy from the incoming planetesimal to Earth's surface. A projectile's energy of motion increases only linearly with its mass but as the square of its velocity, so surprisingly large craters result from relatively small bodies traveling at hypervelocities. Depending on the directions of motion of Earth and of the planetesimal, impacts on the planet may involve relative velocities as high as 50 kilometers per second. At velocities surpassing 4 kilometers per second, the energy of the shock wave created by the impact is far greater than the strength of molecular adhesion for either the planetesimal or Earth. Therefore, on impact the planetesimal acquires the properties of a highly compressed gas and explodes with a force equivalent to a similar mass of blasting powder.

The shock wave from this explosion intensely compresses the target material and causes it to be severely deformed, melted, or even vaporized. In all but the smaller impacts, the entire projectile is also vaporized. The shock wave swiftly expands in a radial fashion, pulverizing the target material and intensely altering the nature of the target rock by extreme and almost instantaneous heat and pressure. This is immediately followed by decompression and what is called a rarefaction wave that restores the ambient pressure. The rarefaction wave moves only over free surfaces, so it travels outward over the ground surface and into the atmosphere above the impact. It becomes the excavating force that lifts vast quantities of the pulverized target material upward and outward to create the crater cavity.

The rarefaction wave excavates a hole whose depth is one-third of its diameter and whose profile follows a parabolic curve, but this depression is short-lived and is therefore called the transient cavity. After passage of the rarefaction wave, a large amount of pulverized target material from the walls of the transient cavity slumps inward under gravity, and some of the ejecta lofted straight up into the atmosphere falls back into the excavation. Together, these sources contribute to a lens-shaped region of breccia that fills the true crater's floor and leaves a shallower, flat-floored apparent crater as the visible scar of the impact. Apparent craters generally exhibit a depth of only one-tenth to one-twentieth of their diameters. Meanwhile, the rarefaction wave carries ejecta particles outward over the surrounding landscape, where they fall to Earth as a blanket of regolith that is distinguishable from the local target rock by the effects of shock metamorphism.

Methods of Study

Impact phenomena are rare enough on the human timescale that no crater-forming events are known to have occurred in recorded history. Owing to this passage of time and to the fact that most existing astroblemes have been severely altered by erosion, impact cratering has been studied by the unique modifications that a powerful impact shock makes in the rocks and

minerals at the site. Scientists study the deformation and structural damage to buried strata, and by looking for the presence of certain rare elements and minerals in the sediments surrounding suspected impact sites.

Much attention has been given to the effects of the shock wave on terrestrial rocks, since shock metamorphism is considered to be the most enduring and positive identifier of ancient astroblemes. Shock metamorphism differs from endogenic metamorphism by the scales of pressure and temperature involved and by the very short duration of the exposure to those pressures and temperatures. Endogenic metamorphism usually involves pressures of less than 1 gigapascal (100,000 atmospheres) and temperatures not greater than 1,250 kelvins. The pressures involved in shock metamorphism are exponentially greater, reaching several hundred gigapascals for an instant in the vicinity of the impact. Rock exposed to pressures in excess of 80 gigapascals and temperatures of several thousand kelvins is immediately vaporized. Lesser pressures and temperatures at increased distances from the point of impact produce signs of melting,

Earth's Largest Impact Craters

Diameter (km)	Location	Crater Name	Age (millions, of years)
300	South Africa	Vredefort	2,023
250	Canada	Sudbury	1,850
170	Mexico	Chicxulub	65
100	Canada	Manicougan	214
100	Russia	Popigai	35
90	Australia	Acraman	590
90	United States	Chesapeake Bay	36
80	Russia	Puchezh-Katunki	175
70	South Africa	Morokweng	145
65	Russia	Kara	73
60	United States	Beaverhead	600
55	Australia	Tookoonooka	128
54	Canada	Charlevoix	357
52	Sweden	Siljan	368
52	Tajikstan	Kara-Kul	5

Source: Data are from the National Aeronautics and Space Administration/Goddard Space Flight Center, National Space Science Data Center.

thermal decomposition, phase transitions, and plastic deformation.

Pockets of melt glass up to several meters thick are commonly found in the breccia within the crater, indicating that pressures there reached 45-60 gigapascals. Coesite and its denser relative, stishovite, are forms of quartz that occur naturally only at impact sites. Shatter cones, conically shaped crystals created at pressures of from 2 to 25 gigapascals, are another prominent feature of shock metamorphism and are particularly well developed in fine-grained isotropic rock. Microscopic examination of impact-shocked porous rock reveals that quartz grains are deformed so as to fill the pores and interlock like the pieces of a jigsaw puzzle. Even at a considerable distance from the impact point, quartz grains tend to be elongated in the direction of the shock wave's passage.

Theories concerning cratering dynamics can also be tested by analogy to some of the craters produced by the detonation of nuclear devices. This latter technique has adequately explained the morphology of the smaller astroblemes, those with diameters that do not exceed 2-4 kilometers. Larger impact events involve additional dynamics that are not mimicked by nuclear devices thus far tested. Astroblemes greater than 2 kilometers in diameter in sedimentary rock or 4 kilometers in diameter in crystalline rock display a pronounced central uplift owing to an intense vertical displacement of the strata under the center of the impact. An additional feature distinguishing complex craters is that their depths are always a much smaller fraction of their diameters than is the case with simple craters.

Photographic imaging of Earth from space has revealed some young and well-preserved astroblemes in remote and poorly explored areas of Earth, such as the Sahara Desert. More important has been the satellite's ability to reveal structures that still preserve a faint but distinct circularity when seen from orbit, although at ground level they are so eroded that their circularity has escaped detection. One of the largest astroblemes yet discovered was detected from Landsat satellite images in this way. New imaging technologies, including advanced radar and sonar mapping, promise to extend the capabilities of space surveillance and remote sensing in recognizing possible impact sites.

Context

The degree to which the Earth is in danger of being struck by a massive planetesimal began to be appreciated about the middle of the twentieth century. In 1980, a team led by Nobel Prize-winning physicist Luis Alvarez announced dramatic evidence suggesting that an asteroid impact that occurred 65 million years ago created such planetary stress that it explained a mysterious massive extinction of life-forms known to have occurred on the Earth at that time. At sites all around the world, researchers discovered that clay deposits at the boundary layer between the Cretaceous and Tertiary periods contained up to one hundred times the normal abundance of the metal iridium, which is rare in Earth's crustal rocks but 1,000 to 10,000 times more abundant in the makeup of many asteroids. This Cretaceous-Tertiary boundary layer is coincident with the point at which fully 70 percent of the life-forms

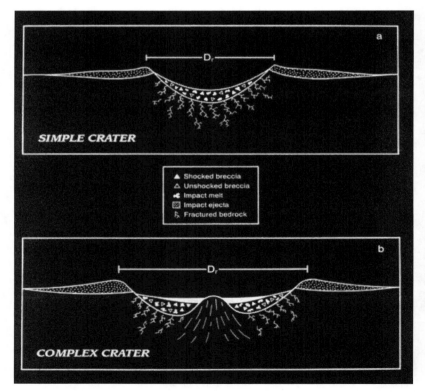

These diagrams show cross sections of the structures of both simple and complex craters. (NASA)

then existing on the Earth, including the dinosaurs, became extinct. Further study has also revealed that this same sediment layer is rich in shock-metamorphosed quartz grains, known only to occur naturally from impact explosions.

Debate continues as to whether an asteroid impact was the primary cause of the mass extinctions at the close of the Cretaceous period or merely the final factor, but there is general agreement that a colossal impact occurred at that time. The volume of material represented in the boundary sediments suggests that the planetesimal was perhaps 10 kilometers in diameter and would have created a crater of as much as 200 kilometers in width. An astrobleme in the Gulf of Mexico near Belize, called the Chicxulub Crater, closely fulfills these criteria. Many scientists accept it as the impact site for the K-T (German for Cretaceous-Tertiary) event. Meanwhile, several other iridium spikes (abnormally high concentrations of the metal) have been found in the sedimentary beds coinciding with other recognized mass extinctions.

Early in the twenty-first century several researchers put forward candidate craters to mark an impact event dated to the time of what paleontologists often call the Great Dying. At the boundary between the Permian and the Triassic (the P-T boundary), which also marks the end of the Paleozoic and the start of the Mesozoic era, life on Earth was very nearly exterminated. A conservative estimate is that 95 percent of all species died out at that time. Life rebounded and the dinosaurs went on to rule the Earth, until they too were wiped out catastrophically. Of the various craters proposed to have resulted from a P-T boundary impact event, the one that appears most likely to turn out to be correct (if any of them are correct) is a crater located in Antarctica, buried unfortunately under 1.5 kilometers of ice. What provides extra confidence that this crater could be the result of a P-T boundary impact event is the fact that at its antipode is located the Siberian Traps. Energy from the impact would have undergone antipodal focusing through the Earth's core to ravage the area on the planet 180° away from the impact site. The Siberian Traps experienced tremendous amounts of volcanic activity around 248 million years ago, the very time of the P-T boundary and the Great Dying. This scenario remains controversial but, if true, would represent an even larger impact event than the accepted K-T boundary event that gave rise to the Chicxulub Crater.

Three related discoveries suggest the possibility that impact cratering may not be an entirely random process, so far as its distribution through time is concerned.

Paleontologists David Raup and J. John Sepkoski, Jr. have shown evidence, based on a rigorous analysis of the marine fossil record, that mass extinctions appear to occur with regularity every 26 million years. Independently, the team of Walter Alvarez (also a member of the team that discovered the K-T iridium anomaly) and Richard Muller have discovered evidence that the ages of major known terrestrial astroblemes seem to be periodically distributed at intervals of roughly 28 million years. For some time, researchers have sought a mechanism that could account for the numerous polarity reversals in Earth's magnetic field over geologic history, and some have suggested that major impact events may be the cause. Several studies have reported an apparent fine-scale periodicity in Earth's magnetic field reversals with a cycle of 30 million years. Although the intervals are not in perfect agreement, they are very close, considering the difficulty of precisely dating extinctions and the exact ages of astroblemes.

These discoveries suggest that there may be an as yet undiscovered member of the solar system that moves in such a way as periodically to disrupt the Oort Cloud, the cloud of comets believed to exist on the fringes of the solar system, causing a barrage of planetesimals to descend upon the inner planets. Although the existence and location of such a body remain speculative and controversial, it has been characterized as a dwarf companion star of the Sun and is called Nemesis.

Richard S. Knapp

Further Reading

Consolmagno, Guy. *Worlds Apart: A Textbook in Planetary Sciences*. Englewood Cliffs, N.J.: Prentice Hall, 1994. A text accessible to college-level science and nonscience majors alike. Presents most topics using low-level mathematics; involves integral calculus where required. Demonstrates how the area of planetary science progresses by questioning previous understandings in the light of new observations.

De Pater, Imke, and Jack J. Lissauer. *Planetary Sciences*. New York: Cambridge University Press, 2001. A challenging and thorough text for students of planetary geology. Covers extrasolar planets and provides an in-depth, contemporary explanation of solar-system formation and evolution. An excellent reference for the most serious reader with a strong science background.

Dixon, Dougal. *The Practical Geologist: The Introductory Guide to the Basics of Geology and to Collecting and Identifying Rocks*. New York: Fireside, 1992. A beginner's guide to the physical processes that formed the Earth and modified its surface over geologic time. Heavily illustrated with guides for rock and mineral identification.

Encrenaz, Thérèse, et al. *The Solar System*. New York: Springer, 2004. A thorough exploration of the solar system from early telescopic observations through the space missions of 2003 that have investigated all planets. Takes an astrophysical approach to place the solar system in a wider context, as just one member of similar systems throughout the universe.

Faure, Gunter, and Teresa M. Mensing. *Introduction to Planetary Science: The Geological Perspective*. New York: Springer, 2007. Designed for college students majoring in Earth sciences, this textbook provides an application of general principles and subject material to bodies throughout the solar system. Excellent for learning comparative planetology.

Grieve, Richard A. F. "Terrestrial Impact Structures." *Annual Review of Earth and Planetary Sciences* 15 (May, 1987): 245-270. A thorough summary of what is known about the cratering process on the Earth, written by a leading authority on the subject. It is intended for the scientific reader, but its illustrations, extensive bibliography, and introductory and summary sections are of value even to those who are not familiar with the concepts and terminology in the body of the article.

Hartmann, William K. "Cratering in the Solar System." *Scientific American* 236 (January, 1977): 84-99. Dated, but a comprehensive explanation of the role attributed to impact cratering in shaping the surfaces of all of the terrestrial planets. The author explains the basis for estimating the frequency of impacts for various sizes of planetesimals and the logic behind using crater counts to estimate the ages of planetary surfaces. The article also explains the theory that the first half billion years of solar-system history involved an extremely heavy bombardment of all the inner planets.

_____. *Moons and Planets*. 5th ed. Belmont, Calif.: Thomson Brooks/Cole, 2005. An updated version of a classic text on planetary science. Includes chapters on all planets and their systems. Discusses the role of impact cratering in shaping planets and their satellites.

Kerr, Richard A. "When Disaster Rains Down from the Sky." *Science* 206 (November 16, 1979): 803-804.

Taking a descriptive approach easily comprehended by laypersons, this article summarizes research by several investigators attempting to compute the frequency with which the Earth is struck by crater-forming bodies. The article places particular emphasis on the Apollo asteroid group and examines suggestions that the Apollo family is supplied with new planetesimals by the decay of former comets.

Morrison, David, and Tobias Owen. *The Planetary System*. 3d ed. San Francisco: Pearson/Addison-Wesley, 2003. Planetary atmospheres are treated as important physical features of the various members of the Sun's family. They are discussed individually in the context of what is known about each planet's characteristics and with regard to theories about their evolution and the evolution of the entire solar system. Geared for the undergraduate college student.

Muller, Richard. *Nemesis: The Death Star*. New York: Weidenfeld & Nicolson, 1988. Despite its tabloid title, this is an excellent discussion of the chain of discoveries leading to the Nemesis theory by the Berkeley physicist who developed it. The volume is organized in two parts: The first recaps the evidence for a major impact at the K-T boundary, and the second tells how further research led Muller to postulate the existence of Nemesis. Intended for lay readers, the book gives insight into how the scientific discovery process works, as well as explaining the theory.

Murray, Bruce, Michael C. Malin, and Ronald Greeley. *Earthlike Planets*. San Francisco: W. H. Freeman, 1981. Although terrestrial impact craters are not specifically discussed, the impact mechanics that produce craters are presented here in terms that are suitable for general readers. A somewhat dated but nevertheless excellent discussion of cratering as a ubiquitous aspect of the surfaces of all the inner planets.

Raup, David M. *The Nemesis Affair*. New York: W. W. Norton, 1986. The author is a significant figure in the field of paleontology and has done leading research on the apparent periodicity of extinctions and magnetic reversals. His narrative is a fascinating personal account of the ideas and the individuals who led the scientific community from extreme skepticism to general acceptance that impact "catastrophism" may have played a major role in the Earth's evolution and its life-forms.

KUIPER BELT

Categories: Small Bodies; The Solar System as a Whole

The observation of object 1992 QB1 led to the discovery of a vast, previously unknown region of the solar system, the Kuiper Belt, which is thought to be the source of most short-period comets.

Overview

Two lines of evidence led scientists to suppose that the Kuiper Belt exists. First, comets that come close enough to the Sun to grow spectacular tails are melting, and so they cannot last forever. Furthermore, if they approach too close to the Sun (or the gas giant Jupiter), gravity may pull them apart, as it did Comet Shoemaker-Levy 9, whose life ended when it impacted Jupiter in 1994. A comet can survive only a limited number of close approaches to the Sun, so why, scientists wondered, were there still so many comets? Where did they come from?

The basic answer is the Oort Cloud, a hollow, spherical cloud of comets with the Sun at its center, extending nearly halfway to neighboring stars. It is generally accepted that long-period comets (those with periods greater than two hundred years) come from the Oort Cloud, and that, if they come close enough to Jupiter to be affected by Jupiter's gravity, some can become short-period comets (those with periods less than two hundred years).

Another feature of comets is that long-period comets come toward the Sun from all directions; that is, their orbital planes may make large angles with the ecliptic plane, the plane of the Earth's orbit around the Sun. It is called the ecliptic because when the Moon is also in this plane, an eclipse can occur. The orbits of long-period comets can be explained if they come from the Oort Cloud, and if the Oort Cloud is spherical. The orbits of short-period comets, however, are more nearly in the ecliptic plane, so their source ought therefore to lie more nearly in the ecliptic plane.

A second line of evidence for the Kuiper Belt comes from modeling the formation of the solar system. There should have been 40 or 50 Earth masses of icy-rocky material beyond Neptune. What happened to that material? In 1949, Kenneth Edgeworth speculated that a reservoir of comets existed beyond the orbit of Neptune. Then, in 1951, Gerard Kuiper reasoned that there should have been large amounts of icy material beyond Pluto and that Pluto's gravity should stir up the icy bodies that formed there, flinging some of them sunward as short-period comets and flinging others into interstellar space. It was thought then that Pluto's mass was far larger than it actually is.

In 1980 Julio Fernández published a paper that used the words "Kuiper" and "comet belt" for the first time in one sentence. Subsequent researchers combined the words, and the term "Kuiper Belt" was born. According to Fernández, a belt of potential comets lies beyond the orbit of Neptune in the ecliptic plane. This belt provides the short-period comets and repopulates the Oort Cloud. Fernández estimated that the Oort Cloud lost three hundred comets per year as they were flung inward toward the Sun or outward to interstellar space. Therefore, the Oort Cloud could not endure for the life of the solar system unless it was being repopulated.

Pluto's strange properties offer another clue that there ought to be a Kuiper Belt. In 1977 James Christy discovered the large satellite of Pluto subsequently named Charon. Using Charon's orbital period and its separation from Pluto, scientists could calculate the masses of both Pluto and Charon. Pluto's mass is 0.0021 Earth mass, only one-sixth the mass of Earth's moon. Charon's mass is 0.0003 Earth mass. Pluto's orbit does not lie in or near the ecliptic plane, as do the orbits of the major planets. Charon was most likely formed in a collision between Pluto and another object perhaps one-fifth the size of Pluto. For this to have any likelihood, there must have been many such objects present, far more than seem to be around today.

The discovery of the Kuiper Belt object (KBO) 2003 UB_{313}, later named Eris, caused astronomers to reevaluate the definition of a planet, since Eris is about 25 percent larger than Pluto. A parallel situation occurred when the first asteroids were discovered and included in the list of planets until it was decided that they belonged to a different class of objects. Likewise, Pluto is not a planet, but the first of the KBOs to be discovered. Pluto and Eris are the largest of the known KBOs and are now classed as dwarf planets, along with the largest asteroid, Ceres.

Also in 1977, Charles Kowal discovered an object that was eventually named Chiron, for the mythological centaur of that name. Chiron is estimated to be 170 kilometers (106 miles) in diameter and orbits between Saturn and Uranus. Apparently Chiron is a comet, because on various occasions it has produced a coma (the large vapor cloud typically identified as the "tail" of a comet). About a hundred such objects have been found with orbits that cross the orbit of at least one of the giant planets. All such objects are referred to as centaurs, and at least three of them have produced comas. Because of gravitational tugs from the giant planets, centaur orbits are unstable over periods of millions of years, so the centaurs now seen must have migrated to their present locations during the past 10 million years. The centaurs will eventually either crash into one of the giant planets or migrate to stabler orbits.

Knowledge Gained

The centaurs and KBOs can be grouped by color: Some are red, and others are blue-gray. The red color can arise as cosmic rays and solar ultraviolet rays strike carbon compounds on an object's icy surface to form reddish-brown compounds. The blue-gray color may arise as smaller objects strike and crater a KBO, covering its surface with new layers of ices. Based on analyses of comets, KBOs must be largely water ice along with carbon dioxide ice, carbon monoxide ice, methane ice, methanol ice, ammonia ice, amorphous (noncrystalline) carbon, silicates and other stony materials, sodium, carbonates, simple hydrocarbons, and clays. The Deep Impact mission found that Comet 9P/Tempel was covered by a dust layer tens of meters deep and the nucleus of the comet was 75 percent empty space, which made it structurally weak. The minor constituents in particular may differ from one comet to another, implying that comets formed at various distances from the Sun.

Experimental evidence of the Kuiper Belt came with the 1992 discovery by Jane Luu and David Jewitt of object 1992 QB_1. It was 41 AU from the Sun, which is about the orbit of Pluto. They estimated its diameter at 250 kilometers. Six months later they found 1993 FW, a second candidate for the Kuiper Belt. By early 2008, more than one thousand KBOs had been discovered. The main belt is shaped like a doughnut centered on the Sun. The doughnut itself goes from 30 AU from the Sun to perhaps 100 AU or even 1,000 AU or more, although there is a sharp falloff in numbers at 50 AU. Objects in the belt are called CKBOs, for "classical Kuiper Belt objects," also called cubewanos, named for the pronunciation of the first CKBO discovered, 1992 QB_1 (Q-B-1-0-s). In order not to be influenced by Neptune at 30 AU from the Sun, these objects must have average distances from the Sun of 40 and 50 AU.

A large fraction of known KBOs are plutinos. Like Pluto, plutinos make two trips around the Sun for every three trips made by Neptune. This guarantees that if they are not now close to Neptune, they will not be close in the future, and therefore their orbits are stable. Since they never get close to Neptune, they may approach to within 30 AU of the Sun with impunity. Pluto never comes closer to Neptune than 17 AU, even though Pluto's orbit crosses Neptune's orbit.

The final group of Kuiper Belt objects are scattered disk objects, or SDOs. Their orbits tend to be more elliptical than the classical Kuiper Belt objects' orbits, and they also travel considerably above and below the ecliptic plane. Scattered disk objects and centaurs seem to form a continuous distribution. If a KBO is ejected to the inner solar system, it becomes a comet or a centaur. It is estimated that there may be 70,000 KBOs between 30 and 50 AU with diameters greater than 100 kilometers (60 miles), and perhaps 1 million with diameters of 1 kilometer (0.6 mile) or larger. There may be as many as 30,000 SDOs that are 100 kilometers (62 miles) in diameter.

Context

It is believed that stars like our Sun form from a more or less spherical cloud of gas and dust within a larger cloud. When conditions are right for one star to form, they are often right for many stars to form; hence, stars, like puppies, tend to be born in litters. The close approach of one or more neighboring stars may have sheared off the Sun's Kuiper Belt at about 50 AU from the Sun. Solid material left over from the formation of the Sun would have developed into a rotating disk with the newborn Sun at its center. Concentrations in the disk formed into asteroid-sized bodies that in turn combined to form the planets. Icy asteroids would have formed among the giant planets and beyond. Those beyond Neptune became the Kuiper Belt. In the inner belt, from about 30 AU to perhaps 50 AU, collisions occasionally occurred, with the result that large objects grew larger (such as Pluto and Eris) and small objects were eventually ground to dust. Beyond 50 AU, objects (if they exist) should be pristine samples of the solar nebula, since they would be so far apart that collisions would be unlikely.

As the newly formed planets interacted with icy asteroids around them, some icy bodies would have been flung inward toward the Sun, and some of these may have hit the Earth and formed the oceans; the Earth was probably born with far less water than it now has. Others were flung outward to form the Oort Cloud and perhaps scattered disk objects. Interactions between KBOs today should occasionally propel them out into the Oort Cloud or inward as comets in the inner solar system or as centaurs in the outer solar system.

Some questions about Pluto and other KBOs may be answered by the New Horizons mission, sponsored by the National Aeronautics and Space Administration (NASA). Launched in 1995, it should reach Pluto by 2015. After taking images in the visible, infrared, and ultraviolet bands of Pluto and its satellite Charon, the spacecraft will go on to one or more KBOs. The infrared images should reveal the surface composition of these objects.

Charles W. Rogers

Further Reading

Davies, John. *Beyond Pluto: Exploring the Outer Limits of the Solar System*. New York: Cambridge University Press, 2001. Excellent and easy-to-read discussion of the ideas and observational evidence leading up to the discovery of the Kuiper Belt. Describes the properties of the various classes of objects found.

Freedman, Roger A., and William J. Kaufmann III. *Universe*. 8th ed. New York: W. H. Freeman, 2008. College-level introductory text covering the field of astronomy. Contains descriptions of astrophysical questions and their relationships.

Lin, Douglas N. C. "The Chaotic Genesis of Planets." *Scientific American* 278, no. 5 (May, 2008): 50-59. Discusses the formation of the solar system, including the formation of the Kuiper Belt and the Oort Cloud.

Luu, Jane X., and David C. Jewitt. "The Kuiper Belt." *Scientific American* 274, no. 5 (May, 1996): 46-52. This landmark article is accessible to the layperson. Luu and Jewitt discuss how they found KBOs and helped to establish the existence of the Kuiper Belt.

Malhotra, Renu. "Migrating Planets." *Scientific American* 281, no. 3 (September, 1999): 56-63. During the formation of the solar system, Jupiter migrated inward, while Saturn, Uranus, and Neptune migrated outward by slinging icy bodies to the inner solar system and to the Kuiper Belt and Oort Cloud.

Stern, S. Alan. "Into the Outer Limits." *Astronomy* 28, no. 9 (September, 2000): 52-55. Discusses history of the Kuiper Belt and how Pluto fits as a KBO.

LUNAR CRATERS

Category: Natural Planetary Satellites

Most lunar craters are the erosion scars of debris left over from the origin of the solar system colliding at high velocities with the surface of the Moon. Studies of sizes and time distributions of lunar impact craters allow scientists to make estimates of the same process acting on the Earth, where much of the evidence has been removed by erosion. Volcanic craters enable researchers to determine the eruption characteristics and thermal evolution of the Moon.

Overview

All of the large lunar craters are named; many of these names are attributable to a 1651 publication by Giambattista Riccioli in which they appear on a map drawn by P. Grimaldi. Riccioli divided the nearside of the Moon into octants. This map was drawn with the aid of the "Galilean" rather than the "astronomical" telescope and so was not inverted. On this map, Octant 1 extended from the ten o'clock position to just past eleven o'clock and was succeeded clockwise by the seven other octants. Craters were named for astronomers, beginning with the most ancient in Octant 1 and concluding with Riccioli's contemporaries in Octant 8. This practice has continued to the present, with the restriction that a crater is always named for a scientist no longer living.

Three principal processes have created the lunar craters. There are those directly excavated by the impact of a meteorite; there are those, called secondaries, that result from the impact of material excavated to form the crater of the primary meteorite impact; and there are those of volcanic origin. Until the return of lunar samples from the Apollo missions (1969-1972), the scientific community had been sharply divided into those who believed the majority of lunar craters to be of impact origin and those who believed the majority to be volcanic in origin. Evidence gained as a result of the Apollo missions has established that the vast majority of lunar craters are of impact origin, resulting from collisions of the Moon with meteors, asteroids, comets, and minor planets (large objects that failed to achieve independent planetary status) at velocities of from 5 to 50 kilometers per second (10,000 to 100,000 miles per hour).

No one has counted the total number of lunar craters, as they range in size from the microscopic to the giant (2,500-kilometer-diameter) South Pole-Aitken basin. It has been estimated, however, that there are about 1,850,000 craters with diameters in excess of 1 kilometer on the lunar surface and 125 with diameters greater than 100 kilometers. A 3,200-kilometer-diameter Procellarum basin has been tentatively identified which, if placed over a map of the United States, would stretch from Washington, D.C., to western Utah and from Brownsville, south Texas, up into central Canada. At the other end of the scale, microscopic impact craters are produced on the Moon because of the lack of a lunar atmosphere. Similar-sized particles rapidly burn up in the Earth's atmosphere.

Primary impact craters increase in morphological complexity with increasing size. Small craters are bowl-shaped, with a well-defined, generally circular rim, smooth interior walls, and a depth-to-diameter ratio of 1:5 to 1:6. The floor of a fresh crater is invariably at a lower elevation than the preexisting terrain. The rim of the crater is surrounded by a generally circular continuous ejecta blanket, followed outward by the discontinuous

ejecta blanket. This discontinuous ejecta often takes the form of rays that radiate outward from a zone close to the center of the primary impact site. An exception is found in craters produced by highly oblique impacts. These craters are generally elongated and have ejecta blankets preferentially distributed downrange or exhibiting a bilateral symmetry, with "wings" on either side of the crater.

An abrupt change in the crater's shape takes place at a diameter of about 16 kilometers in the maria (the Moon's dark lava expanses) and 21 kilometers in the highlands. At larger diameters, craters develop terraces on the interior walls, have a generally broad, level floor interrupted by small hills and mounds, develop a central peak, have a less uniform rim elevation, and have a depth-to-diameter ratio reduced to about 1:40. Flows and ponds, which are often seen both within and exterior to these craters, are impact melts resulting from liquefaction of the impactor and target rocks. At diameters in excess of 140 kilometers, the central peak becomes modified into a centralized peak ring. At diameters in excess of 350 kilometers, multiple rings of alternating elevated and depressed terrains, the giant multiringed basins, are witnessed.

Secondary impact craters are generally less regular than primaries because they are formed at lower collisional velocities, an upper limit being the lunar escape velocity of 2.4 kilometers per second, at which speed objects ejected from the surface would leave the Moon's gravitational field. The size of the secondaries is largely dependent on the size of the primary. Large secondaries have diameters between 2 and 5 percent of that of the primary. Generally, secondaries have smooth interior profiles and are shallower than primaries of the same diameter. They also differ from primaries in that their distribution is nonrandom. They frequently occur in linear or curving chains, patches, or clusters surrounding the primary. Another common feature of secondaries is the presence of a herringbone pattern produced by small ridges ploughed up by impacting objects closely spaced in both time and distance. The apex of the V shape points back toward the primary.

The Galileo spacecraft caught this image of the Moon in 1992, on its way to meet Jupiter in 1995-1997. Major features include Mare Imbrium (center left, Mare Seranitatis and Mare Tranquillitatis (center), Mare Crisium (right edge), and the bright Tycho basin (bottom). (NASA)

Many large lunar craters were once considered to be analogous to terrestrial calderas. Calderas form as a result of collapse following evacuation of a large, near-surface magma chamber. Analyses of returned lunar volcanic materials established, however, that they were derived from great depths (150-400 kilometers), with little evidence of residence at shallow levels for any extended periods. True lunar volcanic craters are recognized primarily on the basis of their distribution, which, like that of secondaries, is nonrandom. Volcanic craters or endogenic craters (those of internal origin) are found at the summits of volcanic domes and cones, at the heads of sinuous rilles, or in association with linear fractures. The craters are generally small (less than 20 kilometers in diameter) and have outlines that range from circular to elliptical to highly irregular. Some volcanic craters are surrounded by a halo of dark surface deposits believed to consist of pyroclastic materials ejected during strombolian- or vulcanian-style eruptions.

Impact craters are the product of an instantaneous geologic event, yet the lunar surface has been subjected to the formation of these features for at least the last 4.2 billion years. Many lunar craters have thus become highly modified from their original pristine form. Since the Moon lacks an atmosphere, this modification results primarily from two agents: later impacts and volcanism. An impact has two principal effects. First, it will result in the total or partial obliteration of any crater smaller than itself that was located within the area of the younger crater. Second, ejecta from the younger crater will erode the walls and infill the floors of craters surrounding itself. At the extreme, the ejecta deposits could totally infill the preexisting surrounding craters. The net result of this process is that older craters are shallower than are newer ones of similar size. The effects of volcanism on impact craters are largely restricted to areas around and within the major maria. A commonly held misconception is that an impact event can trigger the release of magma from the lunar interior. Although the relationship between depth and diameter of the large basins is a subject of much debate, there are very few who believe that the original impact cavities extended to a depth greater than that of the lunar crustal thickness of 75 kilometers. Moreover, there is good evidence that volcanism within any single large basin extended over a time frame of several hundred million years—which is difficult to reconcile with an instantaneous impact event. The reason for the association of volcanic material with impact craters is that the crater floors are topographic

lows and closer to the part of the interior from which the volcanic materials were derived.

One feature attributed to volcanic modification of impact craters is the presence of floor fractures. These features are primarily found on the level floors of craters with diameters of 30 to about 100 kilometers and consist of radial and concentric arrangements of fractures resembling spiderwebs. They are attributed to uplift of the crater floor by subfloor intrusions of magma. Some of these magma bodies found outlets to the surface and, with limited volcanic output, resulted in the formation of volcanic dark-halo craters aligned along the floor fractures. The crater Alphonsus is a typical example. With more extensive volcanism, the floor of the crater becomes flooded with lava until even the central peak becomes buried. At this point, there is too thick an overlying lava pile, and the magma seeks alternative routes to the surface around the periphery of the crater. Some craters have experienced postflood floor fracturing, which results from the sinking of the dense, thick lava pile or posteruptive intrusion, leading to renewed uplift. Flooding of the larger basins appears to have begun within the central low and later extended to the topographic lows between the mountain rings. Small impact craters were constantly being created during the period of basin filling. Many of these craters were either partially or totally covered by younger lava flows.

Methods of Study

Lunar crater studies began in earnest with the availability of the first crude telescopes. At this point it was realized that the Moon's surface was not perfectly smooth, as the Greeks had hypothesized, but instead was pockmarked with features that ranged in size from depressions barely resolved in the telescope to enormous basins. During this telescopic era of investigation, most scientists believed that lunar craters were formed volcanically. However, Eugene Shoemaker of the U.S. Geological Survey became a champion of an impact origin for the majority of the Moon's craters about the time that the space age dawned and early probes began to be sent to investigate the Moon at close range.

Early Pioneer probes (from the late 1950's to the early 1960's) largely failed to achieve lunar goals, but Soviet Luna and American Ranger probes purposely crash-landed on the Moon in several sites and transmitted images up until nearly the instant of impact. Those pictures revealed that the lunar surface displayed cratering down to the smallest scale, providing further evidence for their

*The first close-up of Earth's moon, taken on July 31, 1964, from Ranger 7. The three large craters on the right are (*top to bottom*) Ptolemaeus (*obscured in shadow*), Alphonsus, and Arzachel. Mare Imbrium, pockmarked with smaller craters, fills most of the screen from center to left.* (NASA/JPL)

impact origin, with an enormous number of secondary craters formed out of each primary impact. Several U.S. Surveyor spacecraft (in the mid-1960's) and Soviet Luna spacecraft (from the mid-1960's to 1976) soft-landed on the lunar surface to provide far more images of the nature of the heavily cratered surface, even in the mare regions that appear relatively smooth from backyard telescope views. The Surveyors were followed by Lunar Orbiter and more Soviet Luna spacecraft, which provided detailed maps of cratering across the lunar surface, thereby assisting scientists in both the United States and the Soviet Union to identify safe landing sites to which they could dispatch astronauts and cosmonauts.

Although the Soviet Moon landing program failed, between July, 1969, and December, 1972, six Apollo missions landed at six different sites, returning samples that bore evidence of both impact origin (the breccias) and volcanism (the igneous rocks). In 1976, Luna 24 returned samples to Russia robotically. Then interest in

directly investigating the Moon waned for nearly two decades.

A nearly two-decade drought in lunar exploration ended with the Clementine spacecraft, launched on January 25, 1994, by the National Aeronautics and Space Administration (NASA) and the Ballistic Missile Defense Organization. Although principally a test of new sensor technologies and spacecraft systems, Clementine used the Moon as its target and provided new insights into the Moon's surface distribution of chemicals and minerals. Clementine examined the lunar surface in several different bands of the electromagnetic spectrum, created a laser altimeter-generated map of lunar stratigraphy, and provided a more detailed gravity map of the Moon before its mission ended in June, 1994. Clementine's most exciting finding was that there were likely to be deposits of water ice on the Moon inside permanently shadowed craters, enough water to support human outposts with both potable water and fuel (by breaking down water into hydrogen [fuel] and oxygen [oxidizer]). Clementine's water detection was indirect, in that its sensors picked up the presence of hydrogen (protons), which was then interpreted to be bound in water ice.

Clementine was followed by Lunar Prospector, which launched on January 7, 1998, and conducted a 570-day examination of the lunar surface with alpha particle, neutron, and gamma-ray spectrometers. Lunar Prospector verified the signature seen by Clementine and provided a better estimate of the amount of water ice that might be available in shadowed craters on the Moon. Lunar Prospector was directed to impact the lunar surface on July 31, 1999, in an area (the shadowed crater Shoemaker) where water ice was expected to be found. It was hoped that the impact process might liberate the water ice so that it could be detected from Earth. The final experiment was disappointing: No water plume was seen.

International interest in the Moon increased in the early twenty-first century. The European Space Agency sent the Small Missions for Advanced Research in Technology 1 (SMART 1) spacecraft to the Moon using an advanced ion engine that took thirteen months to reach lunar orbit. From orbit, SMART 1 examined the surface with X-ray and infrared sensors to search for frozen water near the Moon's south pole. Until the mission ended on September 3, 2006, SMART 1 also provided high-resolution optical images of the entire lunar surface, not just craters where scientific interest in ice deposits was centered. SMART 1's mission ended with a purposeful impact at 2 kilometers per second in another attempt to kick up a water plume; results were not definite, although an impact flash was observed from Earth.

Japan launched Selene on September 14, 2007. Two weeks later it entered a highly elliptical initial polar orbit, which was later adjusted to a circular orbit just 100 kilometers above the surface. Its mission was to help set the stage for returning humans to the Moon. At this point NASA, the Chinese, Russians, Japanese, and Indians had expressed, with varying degrees of commitment, plans to land humans on the Moon to establish a permanent outpost. NASA's plans originated with the Bush White House in the aftermath of the *Columbia* accident. George W. Bush announced in early 2004 a Space Vision for Exploration policy which directed NASA to complete the International Space Station by 2010 and then retire the shuttle fleet in favor of returning astronauts to exploration beyond low-Earth orbit. The plan calls for returning astronauts to the Moon to stay by about 2020, and eventually sending humans to Mars and beyond.

The Chinese stated that the goal of its fledgling crewed Shenzhou space program was to send their taikonauts (the Chinese term for astronauts) to the Moon. As a first step the Chang'e 1 spacecraft was launched on October 24, 2007, to survey lunar craters near the South Pole that might be suitable for humans to begin constructing a base of operations for exploration of the Moon as a whole. India launched its Chandrayaan probe on October 22, 2008, entering lunar orbit on November 8. NASA was preparing to launch the Lunar Reconnaissance Orbiter in 2009. Most of these probes were outfitted with detectors with sensing capabilities across several sections of the electromagnetic spectrum, from X rays to infrared radiation. Because of the supposed water-ice deposits there, all spacefaring nations that expressed an interest in setting up human lunar operations concentrated on investigating craters near the South Pole.

NASA established the Constellation program in order to send astronauts back to the Moon. Initial flight operations were planned for late 2014 or early 2015, with the initial lunar mission perhaps coming within five years of those dates. If these missions were carried out, a new age of study of the Moon and its craters would begin.

Applications

Because impact craters are instantaneous events, they are superb geologic time markers. Any material on which an impact crater and its ejecta are superposed is older than the crater; any material overlapping the crater or its ejecta is younger. Analysis of these relationships led to the development of the lunar stratigraphic column consisting of five systems. The pre-Nectarian system, comprising all lunar surface features formed prior to excavation of the Nectaris Basin, and succeeding systems defined by formation of the Imbrium Basin and the Eratosthenes and Copernicus craters.

Primary impact crater densities indicate the relative ages of different units on the lunar surface: the more craters, the older the surface. Craters employed in such studies are usually larger than 4 kilometers in diameter, and the densities are obtained by the extremely tedious task of simply counting them on a photograph. Crater density divided by the average crater-production rate gives the approximate absolute age of the surface units. In the case of the Moon, a calibration curve for production rate can be obtained by comparing the radiometric age of samples returned from the landing sites with crater-count statistics of those sites. These data indicate an exponential decline in crater production from about 4 billion years ago to the present. The details of crater production prior to about 4 billion years are the subject of debate, but, because most of the lunar surface postdates this period, the debate is of little relevance to age determinations.

Small craters (less than 3 kilometers in diameter) have also been used for dating purposes. These techniques are based on the fact that morphologies of small craters are modified in a consistent manner with time. One of these, called the D_L method, is based on the interior slope of the crater. As craters become progressively infilled, the length of the shadow cast by the rim decreases. For a given illumination angle, it is therefore possible to define the largest crater within an area that has reached a specified shadow limit. If a crater in another area is wider than the limit, the second area is older.

Predictable depth-diameter relationships of fresh impact craters allow the determination of some of the third-dimensional characteristics of lunar surface features. If a crater has been flooded by a younger lava flow, the extent of departure of that crater from the dimensions of a similarly sized fresh crater can be employed to determine the thickness of the lava (or other material). Effectiveness of this method is limited by the accuracy of the topographic data.

Material forming the lunar highlands has a different composition from that of the maria and results in pronounced spectroscopic differences. By analyzing spectroscopic signatures of ejecta blankets of craters superposed on the lunar maria, it is possible to determine if a crater excavated solely basaltic material or if it penetrated into the crust beneath. The depth-diameter relationship of the crater can then be employed to ascertain the mare thickness.

Mineralogical and geochemical analyses of returned lunar samples have played a large role in scientists' understanding of the physical processes involved during an impact event. Indirectly, lunar craters have also provided information on the deep lunar interior, because the impacts of both natural and human-made objects have generated seismic waves recorded by the Apollo seismic network.

For a long time claims were made by observers of the Moon that flashes of light came from certain areas of its surface, notably some large craters. Prior to the impact of the Ranger 8 spacecraft there, the crater Alphonsus, for example, was an area where some reported seeing flashes of light; their supposition was that Alphonsus had active volcanism. Most astronomers scoffed at the reports, and no serious observations of Alphonsus ever recorded undeniable light flashes. However, in late 2005 a coordinated search by astronomers at NASA began to record hundreds of small flashes of light representative of explosions on the order of a few hundred kilograms of TNT. The early supposition was that the flashes were the result of volcanic activity. Those recognized flashes were actually from impacts of fragments of the extinct comet 2003 EH1, the progenitor of the Quadrantid meteor shower. This observation indicated ongoing alteration of the lunar surface, a process that had long been attributed to the production of the lunar regolith but was never before seen occurring routinely. The influx of material creating microcraters hinted at a threat to astronauts returning to the Moon to stay and establish permanently occupied bases for research.

Context

The heavily cratered surface of the Moon provides a scenario for what was also taking place on Earth at a time for which scientists have no geologic record. They have learned that, as one goes farther back in time, the number of objects that hit the Moon increases exponentially until around 4 billion years ago. The study of lunar craters has influenced the conceptualization that meteorite impacts may have played a role in terrestrial mass extinctions and thereby the evolution of life. It has been suggested that, because the Moon contains so many craters, a large number of age determinations of the impact events could provide information concerning a hypothesized correlation between impact bombardment and cyclic extinctions.

Analyses of the sizes and distribution of volcanic craters have provided data concerning the internal thermal evolution of the Moon and the stress distributions within the upper lunar crust. In addition, study of the morphologies

In the middleground of this image from Apollo 15, the 40-kilometer-diameter crater Aristarchus can be seen to the left of the 35-kilometer-diameter Herodotus. (NASA)

of impact craters allows scientists to determine the effects of impacts within a gravitational field one-sixth that of the Earth and on a body with no atmosphere. Much of the work by Ralph Baldwin in formulating the characteristics of impact craters was based on data from small terrestrial human-made explosions. Scientists are now able to predict fairly accurately the consequences of very large explosions on land, into water, or in space as a result of lunar impact crater studies. Much of Baldwin's original work remains viable even decades after the first lunar landings.

It has been suggested that the permanently shadowed floors of some near-polar craters may be reservoirs for trapped volatiles such as water. Such resources, if present, would play an important role in the location of a crewed lunar base. Furthermore, impact craters have played the major role in forming the lunar regolith. This loosely aggregated material could be mined with limited mechanical processing. Conversely, the myriad craters on the lunar surface pose a major hazard for safe surface travel and will probably result in unavoidable detours for initial crewed expeditions. Both structures and persons would have to be protected against the small impacting bodies that rain onto the lunar surface.

James L. Whitford-Stark

Further Reading

Baldwin, R. A. *The Measure of the Moon*. Chicago: University of Chicago Press, 1963. The first nine chapters outline the characteristics of both lunar and terrestrial, man-made, and natural craters. Written in pre-metric-system times, this classic text requires mathematical (though simple) conversions. Suitable for advanced high school and college students.

Beattie, Donald A. *Taking Science to the Moon: Lunar Experiments and the Apollo Program*. Baltimore: Johns Hopkins University Press, 2003. Explains the science gleaned from the Apollo lunar landings, including the Apollo Lunar Surface Science Experiment Packages (ALSEPs) and their results.

Chaikin, Andrew. *A Man on the Moon: The Voyages of the Apollo Astronauts*. New York: Penguin, 2007. Reissue of one of the most engaging accounts of the Apollo program and the exploration of the Moon. Critically acclaimed. For all ages.

Consolmagno, Guy, and Martha Schaefer. *Worlds* Apart: A Textbook in Planetary Sciences. Englewood Cliffs, N.J.: Prentice Hall, 1994. A text accessible at the college level for science and nonscience readers alike. Presents subjects at low-level mathematics and also involves integral calculus where required. Demonstrates how the area of planetary science progresses by questioning previous understanding in the light of new observations.

Grego, Peter. *The Moon and How to Observe It*. New York: Kindle Books, 2005. For the amateur observer just starting out, this is the guide that will make one's initial steps at backyard astronomy more enjoyable. Information of use to the more skilled observer too. Primarily uses the Moon as its target in explaining how to make and record observations with telescopes and camera systems, both analog and digital.

Hartmann, William K. *Moons and Planets*. 5th ed. Belmont, Calif.: Thomson Brooks/Cole, 2005. An updated version of a classic text that covers all aspects of planetary science. Strong on Earth-Moon geology. Takes a comparative planetology approach instead of presenting individual chapters on each planet in the solar system.

Jefferis, David. *Return to the Moon*. New York: Crabtree Children's Books, 2007. An explanation of contemporary plans to return American astronauts to the Moon as part of Constellation program operations with the Orion Crew Exploration Vehicle and Altair landing craft. Aimed at a younger audience.

Melosh, H. J. *Impact Cratering: A Geologic Process*. New York: Oxford University Press, 1996. A good synthesis of research information related to the impact process. Mathematics separated out in text and in two appendixes. Aimed at graduate student and research audiences.

Schmitt, Harrison J. *Return to the Moon: Exploration, Enterprise, and Energy in the Human Settlement of Space*. New York: Copernicus Books, 2006. A scientific and economic plan for prolonged exploration and exploitation of lunar resources written by the only geologist to land on the Moon and perform field geology and collect documented samples for return to Earth. Talks about helium-3 mining in order to supply future energy production systems on Earth and thereby finance and systematically expand lunar settlements and research posts.

Schrunk, David, Burton Sharpe, Bonnie Li Cooper, and Madhu Thangavelu. *The Moon: Resources, Future Development and Settlement*. New York: Springer Praxis, 2007. Examines what is needed to develop human settlements on the lunar surface beginning with renewed robotic exploration. Explains the benefits of a robust spacefaring civilization.

Spudis, Paul D. *The Geology of Multi-ring Impact Basins: The Moon and Other Planets*. New York: Cambridge University Press, 1993. A comprehensive geological study of large impact craters formed on the planets and their satellites. Extensively illustrated with photography and diagrams. Suitable for both the lay reader and scientific audiences.

Wilhelms, D. E. *The Geologic History of the Moon*. U.S. Geological Survey Professional Paper 1348. Washington, D.C.: Government Printing Office, 1987. Full of photographs of lunar craters supplemented with an outstanding text based on more than twenty years of lunar geological mapping by the author. Research-level book but understandable to the dedicated high school student. A must-have book for those interested in the Moon.

LUNAR HISTORY

Category: Natural Planetary Satellites

The heavily cratered lunar surface has slowly evolved over the past 4.6 billion years primarily because of impact events. Observations made by astronomers with ground-based telescopes, studies carried out a result of the Apollo program, and direct studies of Moon rocks have revealed much about this formation process.

Overview

The Moon's surface has evolved into its present state as a result of meteoritic impact over the eons since the solar system formed. In contrast, Earth's surface has been molded and shaped primarily as a result of geologic activity brought on by heat transfer from its molten core. One can argue that these very different surfaces—the Earth's constantly changing, eroding surface and the Moon's relatively tranquil, slowly changing surface—are the result of each body's total mass. Generally, low-mass bodies acquire surfaces with meager geological activity, whereas more massive bodies continuously undergo a considerable amount of geologic change. There are thought to have been six states in the evolution of the Moon: the origin of the Moon, the separation (or differentiation) of the Moon's crust, the first age of igneous activity, the great bombardment period, the second age of igneous activity, and the quiescent period.

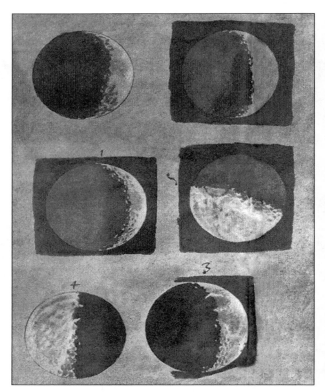

The astronomer Galileo published these sketches of his lunar observations in Sidereus Nuncius in 1610. (The Galileo Project, Rice University)

The origin of the Moon is the stage that is least understood. Astronomers, physicists, geologists, and mathematicians have struggled and debated over the origin of the Moon for at least two centuries. The original theory of the Moon's formation was that it was a body accreted elsewhere within the solar system and came close enough to the Earth to be captured by its gravitational field. However, such a hypothesis proved difficult to model mathematically and was eventually proven to be impossible in a two-body system with masses such as those involved in the Earth-Moon system.

Another hypothesis was that the Earth and the Moon formed about the same time, shortly after the proto-Sun (the very early stage of the Sun, which contracted by gravity out of interstellar material) some 4.6 billion years ago. The Earth and the Moon formed relatively near each other (at a distance that is small compared to the distance between planets). This theory lost favor, however—especially after determination of the chemical composition of Moon rocks. Lunar samples were devoid of many common elements in Earth rocks. Because the Moon lacks water, the number of lunar minerals relative to those found on Earth is quite small.

A modern hypothesis of the lunar origin is called the large impact hypothesis. It gained favor among scientists and then became subject to intense investigation and scrutiny. The idea is that a Mars-sized body struck the early Earth. Such a catastrophic collision would have resulted in the ejection of a large cloud of debris from the obliteration of the Mars-sized object and severe damage to the early Earth. Subsequently, the material would have condensed into a disk of debris, some of which would have fallen back to Earth, completing the formation of the Earth. In time the rest would have accreted into the Moon. This theory has many features that are consistent with the knowledge that the Earth and Moon formed near each other and explains the similarity of the Earth's crust to the composition of the Moon. It also explains why Earth's moon is so much larger in proportion to its parent planet than any other satellite in the solar system.

The second stage of evolution, the differentiation of the crust, occurred during the early stages of origin, when the Moon was still molten in its outer layers. This condition occurred from the heat of formation and was prolonged by constant bombardment of meteorites striking the Moon's outer layers. In this early stage of the solar system, much debris was still in the form of a complex set of rings around the Sun. Chunks of the debris, composed of meteoroids, fell onto the surfaces of the Moon and other planets. These meteoritic showers were especially frequent in the early solar system. Since the outer layers of the Moon were molten, the lighter minerals tended to float and the heavier ones tended to sink. This means that heavy igneous material, such as the mineral basalt, would sink, whereas lighter igneous material, such as the mineral anorthosite, would float. The Moon's crust differentiated into an igneous shell with an inner zone of basalt and an outer zone of the lower-density igneous rock anorthosite. Nearly all specimens collected during the Apollo program can be put into three categories: mare basalt, found in the maria (the basins, from the Latin for "seas"); KREEP-norite, with an unusually high content of potassium (K), rare Earth elements (REE), and phosphorus (P); and the anorthosite group. The latter two are found in the lunar highlands.

As the Moon's surface solidified, several pockets or subsurface pools of KREEP-norite remained molten because of slight differences in chemical composition within the crustal rock. This episode of igneous activity is the third stage of lunar evolution. Certain impurities would form crystals with lower melting temperatures and thus remain liquid. Some of the darker shades of pale-yellow-rock highland material, surrounding the maria, are KREEP. Coincidentally, meteorites struck above some of these liquid pockets and fractured the solid rock. The cracks and fractures formed would occasionally lead to one of these pools, and the release in pressure would cause the liquid to "creep out" onto the surface.

The fourth stage, the great bombardment period, overlaps all the other stages. Its placement in the list of the Moon's evolutionary stages is based on the concept that this age is easily identified by the largest craters still visible today. Continuously during the first 500 million years and the first four stages, impacting bodies of all sizes struck the Moon. The largest formed multirimmed basins, hundreds of kilometers in diameter. The meteoroids responsible may have been huge bodies 150 to 200 kilometers in diameter. Many of the largest craters observed in the highlands date back to impacts during this period. This is also true of many of the large basins, which in turn are often accompanied by the gray-blue colored maria.

As the result of a high-speed impact, both the meteoroid and the underlying surface material of the Moon were subject to rapid, intense heating. All the kinetic energy of the projectile's motion was translated into other forms, namely heat, shock waves, and energy of excavation of lunar material, since the motion of the impacting body was halted abruptly. In a fraction of a second, outer regions of the meteoroid and some surface material were vaporized. The resulting high temperature caused dense gas to explode violently. This activity resulted in a crater surrounded by a rim, composed mostly of the material that was ejected from the resulting hole. Beyond the rim, arranged in a sunburst or spray pattern, was the ejecta blanket, complete with innumerable small craters (called secondaries) of a wide range of sizes.

Since the mass of the Moon is low compared to that of Earth, the Moon lost most of its internal heat of formation and the heat provided by radioactive decay in a short time, perhaps less than 800 million years after its origin. Earth, being eighty-one times more massive than the Moon, has retained much of its heat, as a result of radioactive decay in its interior. Although the Moon may still possess a molten core, it does not transfer a substantial heat flow to the surface, as it once did. Heat-flow experiments performed by Apollo astronauts on the Moon's surface did register an outflow of heat, but those experiments were rather rudimentary. More sophisticated and far-ranging detectors are needed to determine the Moon's internal thermal state more precisely.

Between 3.8 and 3.1 billion years ago—during the fifth stage of lunar history, the second great igneous period—only the lower basaltic strata of the separated crust were heated. The largest impacts, slightly after and during this stage, formed large fractures that in some cases reached down to the lower basaltic layers. This allowed molten basalt to flow as lava to the surface and fill in the low-lying areas in and around the basins. This accounts for the gray-blue basaltic maria. All these maria have generally circular shapes, with one exception, the Ocean of Storms. This pancake shape is what one would expect from flows into large, almost circular craters. Since basalt has a higher density than anorthosite, these pancaked lava flows have been detected by the effect their gravity has on satellites orbiting the Moon. As a result of their localized pull of gravity, they have been named mascons (mass concentrations). The flow of basalt onto the lunar surface apparently came in stages, since astrogeologists can identify younger flows atop older flows. The age of the flows can be identified by comparing the frequency of impact craters found in the flows and also the wrinkle structures in the maria that identify the border of a flow.

Finally, after the first one billion years, the igneous activity ceased and impacts gradually became less frequent. This final stage, the quiescent period, lasted more than three billion years to the present time.

Methods of Study

Almost all the features on the Moon's surface can be shown to be caused by impacts—directly, as in the case of a crater, or indirectly, as in the case of KREEP or a maria. There are no mountains on the Moon resulting from uplifting. All the mountainlike features are partial or quite complete rims of basins or craters. The partial rims have been obliterated by more recent impacts. Another mountainlike feature often seen at the center of a well-preserved or relatively fresh (young) crater results from the focus of a shock wave reflected

and focused by the rim formation during impact back toward the impact center. Arrival of the reflected shock wave apparently heaves the center of the crater upward.

The surface of the Moon today is covered with a finely pulverized dust called regolith. Regolith varies in depth from 5 to 10 meters in young maria (3.1 billion years old) to perhaps 20 to 25 meters or more in the highlands (more than 4.2 billion years old). Regolith has been formed by the constant churning and mixing from particles impacting through the ages, ranging in size from micrometeorites to meteoroids large enough to form the largest craters. Crater rims, having both a sharp (young) and a rounded, smoothed (old) appearance, are evidence of this same kind of slow erosive process caused by impacts.

Close microscopic inspection of the regolith reveals evidence of impacts and the Moon's evolutionary history. The regolith consists of tiny pieces of anorthosite, anorthositic breccia, basalt, basaltic breccia, a variety of glasslike, irregularly shaped particles, and small,

Even in black and white, this false-color image—compiled from 15 different images of the Moon from the Galileo mission in late 1992—highlights the heavily pockmarked surface; false color was used to assist planetary geologists in identifying the types of rock and minerals that dominate different regions. (NASA/JPL)

spherically shaped glass beads. The glass is believed to be produced by the heat and shock of impact, resulting in a fused metamorphic, glasslike rock. Spherical beads result from liquid drops thrown out during the splash of ejection shortly after impact. The drops freeze or solidify before falling back to the surface.

In the early 1970's, Apollo astronauts conducted field geology exercises on the Moon's surface and placed experiments there to study moonquakes in the hope of getting information about the lunar interior. Moon rocks brought back to the Earth played a major role, if not the most important role, in advancing understanding of the evolution of the lunar surface. Fragmented pieces of lunar rock provide many clues to lunar history.

Geologists can easily identify several distinct rock types in such samples. Some of the gray pieces are the well-known igneous rock basalt. Some of the very white pieces are brecciated anorthosite; breccia is rock that has been fragmented and welded together with immense heat. The glasslike substances and those resembling glass beads provide an understanding of the evolution of the Moon's surface as brought about by eons of meteoritic bombardment. Apollo studies, together with previous space-based and earthbound telescopic studies, have allowed scientists to construct an evolutionary history of the lunar surface.

All lunar samples are geologically processed. Their composition was established by igneous processes inside the Moon. No primitive or primordial lunar material (material that existed on the Moon's surface soon after the formation of the Moon) is likely to have survived the turbulent early history of the Moon. Primordial material from which the Moon was formed should be about 4.6 billion years old. The Earth and Moon, moreover, formed in the same general region. This is indicated by the specific abundances of certain types of atoms, especially those of oxygen. Abundances are good indicators of position from the Sun in the inner solar system. Lunar and crustal terrestrial rocks have very similar abundances. The overall density of the Moon is similar to the density of the Earth's crust; for this reason, scientists in the past speculated that the Moon came from the Earth.

Neither humans nor robots have visited the lunar surface and returned samples to Earth since Luna 24 in 1976. The Ballistic Defense Organization used the Moon as a test bed in early 1994 for an evaluation of next-generation sensors flown on the Clementine spacecraft. This small probe, which lasted only 115 days, carried seven different instruments capable of examining the lunar surface across

much of the electromagnetic spectrum. In coordination with the National Aeronautics and Space Administration (NASA) a mineralogical map of much of the Moon was obtained. Surprisingly, there was a signature strongly suggesting that the Moon possessed a significant amount of water ice near the poles. Clementine was followed four years later by NASA's Lunar Prospector, which expanded upon the Clementine science. Over the course of nearly eighteen months, Lunar Prospector orbited the Moon and assembled data that provided a detailed map of surface composition and identified lunar resources that might be used to support lunar bases.

In the aftermath of the *Columbia* accident (February 1, 2003), the Bush administration directed NASA to proceed with a next-generation crewed spacecraft that could again take astronauts beyond low-Earth orbit, expand the human presence to return to the Moon to stay, and eventually fly to Mars and beyond. Rather than setting up a time-limited crash program like Apollo, the Bush administration established the Space Vision for Exploration, which was open-ended. NASA responded to the challenge with the Constellation program, which included development of new boosters to send an Orion Crew Exploration Vehicle to the Moon with a much larger and more versatile Altair Lunar Lander in order to establish a base. The precise site of the base remained under evaluation, but its general vicinity would be near the Moon's south pole to make use of the suspected water ice deposits there and to also be able to take advantage of permanent solar irradiation for power generation. Plans called for a return to the Moon with Constellation assets by 2019, the fiftieth anniversary of the first crewed lunar landing (Apollo 11 in July, 1969). In addition to setting up infrastructure at the Moon's south pole, Constellation was to be designed to support exploration about the lunar surface far from the base itself to attempt to answer many of the questions about the Moon's history and geology that Apollo did not completely answer.

The United States was not the only nation at this point with interest in the Moon. After much indifference, international interest in understanding the Moon's history and making determinations about potential uses of its resources began to increase dramatically in the early twenty-first century. China began sending taikonauts (the Chinese word for astronauts) into space in Shenzhou spacecraft in late 2003. The Chinese crewed space program was designed to incrementally advance to lunar missions. Indeed, it is possible that Chinese taikonauts will reach the Moon before NASA astronauts return.

NASA's next robotic spacecraft dedicated to lunar science is the Lunar Reconnaissance Orbiter (LRO). LRO joins probes sent to the Moon by the European Space Agency (SMART 1 spacecraft), the Japanese Aerospace Exploration Agency (Kaguya spacecraft), and the Indian Space Agency (Chandraayan-1 spacecraft).

Context

Understanding the history of the lunar surface has helped scientists to discern the events and processes that led to the origin of the solar system, the planets in general, and the Earth-Moon system in particular. It is important to understand the history of the Moon in that it provides information about the history of the Earth. Because of the Moon's ancient surface, questions that cannot be answered by examining the youthful terrestrial surface have been answered by lunar studies.

Detailed studies of lunar rocks have led to a much deeper understanding of the probable chemical composition of the solar nebula and the proto-Sun. This is important not only because it helps explain the process whereby the Sun and planets formed but also because it provides specific clues and observations that increase scientists' understanding of star formation in general. Astronomers now believe that planetary formation is a natural consequence of star formation.

Anomalous abundances of particular chemical elements found in both lunar and terrestrial rock have led directly to speculations about the origin of the solar system. Some chemicals in the rare Earth group strongly indicate that a supernova explosion may have contaminated the gas and dust from which the solar nebula formed with chemical debris from deep within the exploding star. Some astronomers suggest that the shock of the supernova blast wave actually triggered the contraction of the solar nebula into the formation of the Sun and planets.

Understanding the lunar surface also advances the goal of its future exploitation. The Moon might someday be used as a space station, an astronomical observatory, or a space colony. The Moon would be an excellent source of minerals, since virtually all lunar rocks are rich in the metal titanium and would supply much more of this metal than could be mined on Earth from even the richest deposits. To these ends, NASA is developing the Constellation program, which includes returning humans to the Moon. The goal of Constellation is to expand the human presence beyond low-Earth orbit. The first major step in that lofty, open-ended adventure includes the development of a permanently staffed lunar base, one from

which exploration and science can be conducted, all the while making use of lunar resources to meet the needs of the scientists and astronauts living and working on the Moon.

James C. LoPresto

Further Reading

Abell, George O., David Morrison, and Sidney C. Wolff. *Exploration of the Universe.* 7th ed. Philadelphia: Saunders College Publishing, 1995. This elementary astronomy textbook is considered by many astronomers to be one of the finest available. It is written in a traditional style, and the chapter on the Moon includes several good photographs, images, and diagrams.

Baldwin, R. A. *The Measure of the Moon.* Chicago: University of Chicago Press, 1963. The first nine chapters outline the characteristics of both lunar and terrestrial, human-made, and natural craters. Written in pre-metric-system times, this classic text requires mathematical (though simple) conversions. Suitable for advanced high school and college students.

Beattie, Donald A. *Taking Science to the Moon: Lunar Experiments and the Apollo Program.* Baltimore: Johns Hopkins University Press, 2003. Explains the science gleaned from the Apollo lunar landings, including the Apollo Lunar Surface Science Experiment Packages (ALSEPs) and their results.

Freedman, Roger A., and William J. Kaufmann III. *Universe.* 8th ed. New York: W. H. Freeman, 2008. College-level introductory text covering the field of astronomy. Contains descriptions of astrophysical questions and their relationships. Informative.

Grego, Peter. *The Moon and How to Observe It.* New York: Kindle Books, 2005. For the beginner or amateur observer, this is the guide that will make one's initial steps at backyard astronomy more enjoyable. Primarily uses the Moon as its target in explaining how to make and record observations with telescopes and camera systems, both analog and digital. Of use to more skilled observers, too.

Hartmann, William K. *Moons and Planets.* 5th ed. Belmont, Calif.: Thomson Brooks/Cole, 2005. An updated version of a classic text that covers all aspects of planetary science. Explains all current theories for the formation of the Moon.

Kosofsky, L. J., and Farouk El-Baz. *The Moon as Viewed by the Lunar Orbiter.* NASA SP-200. Washington, D.C.: Government Printing Office, 1970. A fine collection of early lunar images taken from the Lunar Orbiter satellite series. They consist of both overall global images and detailed high-resolution images of particular areas on the Moon.

Wilhelms, Don E. *The Geologic History of the Moon*. Professional Paper 1348. Denver, Colo.: U.S. Geological Survey, Books and Open-File Reports Section, 1987. A detailed publication on lunar history with a strong emphasis on what has been learned as a result of the space program.

_____. *To a Rocky Moon: A Geologist's History of Lunar Exploration*. Phoenix: University of Arizona Press, 1994. Numerous books portray the Apollo astronauts, who successfully completed six lunar landings, with heroic prose. Others describe the science learned from Apollo investigations and returned Moon rocks and soil samples. This book is both a personal account and a geology-driven explanation for the way individual missions were planned and carried out for maximum science return. Includes descriptions of real-time evaluations of field geology aspects on each Apollo mission.

LUNAR INTERIOR

Category: Natural Planetary Satellites

Using fundamental knowledge of the Earth's interior along with data returned by missions to the Moon, scientists have been able to extrapolate theories about the lunar interior.

Overview

The contrasting light-colored highlands and darker maria regions of the Moon were very apparent to even the earliest observers. Early telescopic observations revealed the highlands to be rough, cratered, and mountainous, as compared to the smoother and less cratered maria. Galileo's first impression of the lunar surface clearly drew a comparison to the land and sea regions of the Earth. Later scientists, with only telescopic observations to guide them, assumed that the Earth and Moon probably had similar origins and should be quite alike in most respects. This assumption did not last long.

When scientists made their first calculations of the densities of the Earth and Moon, they discovered an interesting fact. The Earth has a density of 5.5 compared to 3.34 for the Moon. This seemed strange for two bodies found so close to each other in space. Later theories attributed the difference in densities to the Earth's having a much higher percentage of metal in its overall chemical composition. With the higher metal content, the Earth would naturally have a greater mass and a higher density. Two distinctly different sets of characteristics would soon define each object, for it is mass that creates a planet's internal pressure and temperature conditions. These two factors, in turn, determine whether or not a metal or silicate mineral will remain a solid or become a molten liquid. Once a molten liquid is produced, the process of density separation can take place and produce a distinct core.

Based on a density comparison between the Earth and the Moon, it is believed that the Moon does not have a well-defined core like that of the Earth. Experiments conducted on the Apollo 15 and 17 missions did indicate that the Moon's heat-flow rate is about half that of the Earth. Although the lunar interior is hot, it is not sufficiently hot to produce a density separation of materials comparable to that which occurred in the Earth. The higher iron content of the lunar crust also tends to support a planetary body that is not well differentiated like the Earth.

The current model for the nature and chemical composition of the Moon's interior has been primarily derived from lunar rock samples and the seismic experiments left on the Moon by the Apollo astronauts. This information, combined with data from both crewed and uncrewed orbital missions, has given scientists a much clearer idea of what constitutes the lunar interior. Originally, based on the relatively well-established models developed for the Earth, it was assumed that the Moon's interior should have experienced a similar history. As Apollo data started to pour in, it quickly became apparent that this was not going to be the case. In subsequent years, differences between the interiors of the Earth and Moon have become even more controversial.

The interior of the Earth has been divided into a crust, mantle, and core. This division is based on the determination of the densities of various rocks and metals at specific depths. This division does not represent the primordial Earth. One theory of the origin of the Earth suggests that the Earth formed as a result of the accretion of innumerable cold, solid bits of rock and metal into a relatively cold, chemically homogeneous body. Shortly after the Earth's formation, its internal temperature began to rise due to the decay of radioactive isotopes, the heat from accretion, and the mass of the Earth itself. At this point,

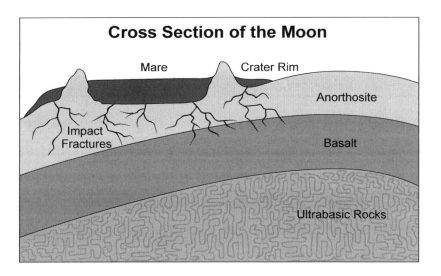

Cross Section of the Moon

Mare Crater Rim

Anorthosite

Impact Fractures

Basalt

Ultrabasic Rocks

its interior temperature was well below the melting point of most metals and silicate minerals. Gradually temperatures reached a point where melting occurred, producing a molten liquid. Within this molten magma, denser metals sank to the center of mass forming the liquid and solid metallic cores. As a result, lower-density silicate minerals were displaced and moved upward to form the crust.

Even within Earth's crust itself, rock materials of two distinctly different densities would separate into the lower-density continental crust and the higher-density oceanic crust. Sandwiched between the crust and core would be the mantle, a large region of silicate minerals with variable densities appropriate to specific depths. This entire process is referred to as density separation, and it is believed to have taken place very early during the initial stages of the Earth's formation. Through this separation of molten metals and nonmetals, the generation of an electric current and magnetic field became possible.

If this process occurred for Earth, then could a similar process have taken place on the Moon? Perhaps the best indicator of the composition and structure of the lunar interior came from the Apollo Lunar Science Experiment Packages (ALSEPs), which included a seismometer to record "moonquakes." On Earth, scientists use seismic waves to calculate the density and predict the composition of materials at various depths. The way seismic waves pass through various materials gives clues to their chemical and physical properties. The same should be true for the Moon and thus provide a detailed picture of the lunar interior. Apollo results indicate that the internal structure of the Moon is very different from that of the

Earth. Both the lunar crust and mantle are much thicker relative to the Earth's and show no evidence of plate tectonics. Seismic evidence does indicate that the upper portion of the lunar crust has been shattered by countless meteoroid impacts; with depth, this crust gradually progresses into a solid rock layer termed the lithosphere. Beneath this first 1,000 kilometers (620 miles) lies the asthenosphere, a region where seismic waves have indicated the presence of a liquid or partially liquid environment, which may include the core. The core, if it truly exists, may consist of an iron sulfide mixture rather than a "pure" nickel-iron alloy. The presence of a significant amount of iron sulfide minerals could lower the melting temperature required to produce a liquid phase and hence the conditions suitable for the formation of a core. The specific chemical composition of the lunar interior, along with the existing temperature and pressure conditions, are the defining factors as to whether or not the Moon has a core similar to Earth's. Existing data suggest that the Moon does not have sufficient internal heat to produce a distinct metallic core. The fact that iron-bearing lunar materials are not magnetic also points to the apparent lack of a magnetic field, which is usually attributed to the presence of a liquid-solid metallic core.

Knowledge Gained

The arrival of the "space age" gave scientists the opportunity to answer many of their questions about the origin and evolution of the planets, because new technology allowed them to send satellites and probes to explore other worlds at close range. Because of its close proximity to Earth, the Moon was a logical first choice for exploration. Several hundred years of telescopic studies had not provided many answers; to unravel the Moon's secrets, both crewed and uncrewed spacecraft would need to go to the Moon and collect data.

Beginning in the early 1960's, when President John F. Kennedy challenged the American people to go to the Moon, the National Aeronautics and Space Administration (NASA) developed a series of lunar exploration programs: the Ranger, Lunar Orbiter, Surveyor, and Apollo missions achieved that goal. Ranger provided the first close-up look at the Moon. Lunar Orbiter provided the reconnaissance images to select the landing sites for Surveyor and later the Apollo missions. In addition to proving that a lunar soft landing was possible, Surveyor gave science its first look at the lunar soil and surface conditions. The six Apollo missions not only returned more than 380 kilograms of lunar rock and soil but also left experiments on the lunar surface to study the Moon's interior. Later missions, such as Clementine and the Lunar Prospector, surveyed the lunar surface for mineral deposits and searched for the presence of water ice. The Russian Luna program also added to our overall knowledge of the Moon through the use of a robotic rover and sample-return missions.

Knowledge seems to flow back and forth between the Earth and Moon. The basic geological principles geologists have learned on Earth have been applied to lunar features, yet many of the discoveries made on the Moon have caused geologists to rethink their original theories. In the context of a wealth of lunar data, scientists now see the Moon as a world seemingly similar to Earth yet markedly different. The early processes that created a distinct crust-mantle-core structure for the Earth and produced a strong magnetic field never reached completion on the Moon. The Earth was able to retain its high internal temperature and remain fluid at specific depths, while the Moon apparently did not and remains only a partially differentiated body. Further studies of the Moon are certainly needed before science can provide a definitive understanding of the Moon's physical makeup and interior structure. Future lunar missions may answer many of the remaining questions concerning the lunar interior as well as giving a better understanding of lunar surface materials and the giant impact processes that have shaped lunar history.

While lunar geologists eagerly await a return to the Moon, analysis of samples returned by the Apollo astronauts continued decades after the last lunar landing, Apollo 17 in December, 1972. A rock returned on that mission after being collected by Dr. Harrison Schmitt, the only geologist to land on the Moon during Apollo, turned out to

be the oldest sample collected that had not been subjected to intense shocks from major bombardments of the Moon occurring after the rock's formation. This rock possessed a remnant magnetism dating back to beyond 4.2 billion years, indicating that in its early history the Moon had a liquid core that produced magnetism by a dynamo effect, as Earth still does. That lunar field appears to be about one-fiftieth that of the Earth, a result that models of lunar core dynamics also predict. This finding adds evidence to the theory that the Moon did not form cold but out of a collision between a Mars-sized object and the early Earth.

Context

As far back as 400 B.C.E., ancient Greek philosophers wondered about what lies beneath the Earth's surface. They envisioned a dark, hot, sulfurous underworld populated by demons and the spirits of the dead. This dismal picture was based on a certain amount of truth. The ancients were familiar with volcanoes and sulfurous hot springs and easily made the connection to the underworld.

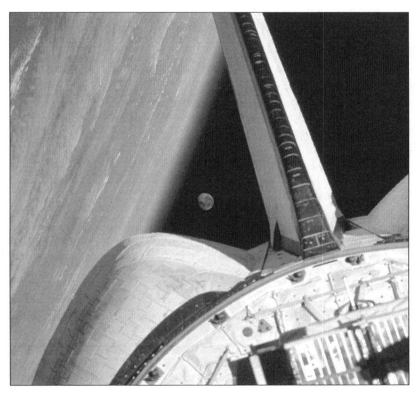

In 1998 astronauts aboard space shuttle Discovery took this picture of the Moon from Earth orbit, as seen to the right of the Atlantic Ocean through the shuttle's aft windows. (NASA)

Today modern science certainly has an advantage over the Greeks. Scientists now have the ability to study the Earth's interior by means of seismic studies, deep drill holes, and analyses of deep-seated rocks brought up during violent volcanic eruptions. With data like these, it is possible to develop computer models that give a very accurate picture of what lies deep beneath the Earth's crust. Once we have a basic idea of what makes our planet work, natural human curiosity causes us to wonder if these same geological processes are common to other worlds. For more than fifty years, humankind has extended its curiosity to our Moon and found a world that is both familiar and different from Earth. Certainly, future lunar missions will not only expand our knowledge of the Moon's interior but also pave the way for a more comprehensive understanding of the terrestrial planets.

Paul P. Sipiera

Further Reading

Canup, R. M., and K. Righter, eds. *Origin of the Earth and Moon*. Tucson: University of Arizona Press, 2000. A compilation of twenty-nine scientific papers dealing with terrestrial planetary formation with emphasis on the Earth-Moon system. Excellent reference source for undergraduate and graduate students.

Hartmann, William K. *Moons and Planets*. 5th ed. Belmont, Calif.: Thomson Brooks/Cole, 2005. A comprehensive examination of our solar system from the planetary geologist's perspective. Suitable for advanced high school students and undergraduates.

Hubbard, William B. *Planetary Interiors*. New York: Van Nostrand Reinhold, 1984. A classic work that addresses the fundamentals of planetary interiors with detailed coverage of the Earth and Moon. Appropriate for high school and college-level readers.

Lang, Kenneth R. *The Cambridge Guide to the Solar System*. Cambridge, England: Cambridge University Press, 2003. A concise yet comprehensive reference for all members of the solar system, with detailed information on the lunar interior. Suitable for a wide range of readers.

Mohit, P. Sundas. "The Two-Faced Moon." *American Scientist* 96, no. 3 (May/June, 2008): 210-217. Addresses the intriguing issue of the Moon's variable crustal thickness with many references to the nature of the lunar interior. Suitable for a general audience.

Mutch, Thomas A. *Geology of the Moon*. Rev. ed. Princeton, N.J.: Princeton University Press, 1972. A classic work written by one of the leading planetary geologists of the Apollo era. Despite its publication date, this work still serves as an excellent starting point for an understanding of Earth's moon. Suitable for the general reader.

Taylor, Stuart Ross. *Planetary Science: A Lunar Perspective*. Houston, Tex.: Lunar and Planetary Institute, 1982. An often-referenced source in the current literature, despite its age. Provides the reader with fundamental information that aids understanding of later developments in lunar science.

LUNAR MARIA

Category: Natural Planetary Satellites

The Moon's maria (literally "seas") are low-elevation areas of the lunar surface, in contrast to the lunar highlands. Interest in lunar maria achieved international prominence in 1969, when the National Aeronautics and Space Administration's (NASA's) Apollo 11 astronaut Neil Armstrong set foot upon the surface of the Moon in the Sea of Tranquility. There Armstrong and fellow astronaut Edwin E. Aldrin collected the first samples of the surface of another natural body in space.

Overview

Mare (plural, *maria*) is the Latin for "sea"; the term reflects seventeenth century ideas about the nature of the lunar surface. Similarly, there are other places on the lunar surface named for bodies of water: the Oceanus Procellarum (Ocean of Storms), some areas bearing the name *lacus* (lake), others called *sinus* (bay), and still others called *palus* (marsh). Returned lunar samples have established not only that there is no surface water on the Moon but also that the Moon as a whole has so little water that there are no hydrated minerals in the rocks. In reality, the maria are areas of basaltic lava flows. These lavas occupy approximately 16 percent of the lunar surface, with 80 percent of those located in the equatorial area on the side of the Moon that is always turned toward the Earth—a total of more than 6 million square kilometers. Although extensive, mare basalts probably represent less than 1 percent of total lunar crust by volume.

Volcanoes form some of the highest features on the Earth, but the majority of the lunar lavas are situated in low regions. These lows were excavated by meteorite impacts that left craters varying in size from microscopic to 2,500 kilometers in diameter. The largest of these craters, called basins, were later filled by lavas to form the maria. Smaller craters, particularly those immediately surrounding the nearside maria, also became the sites of lava eruptions. Estimates of the thicknesses of the mare lavas range from less than 100 meters in the shallowly flooded craters to perhaps 10 kilometers in the larger basins. Depth of basalt fill appears to be related to the age of the impact-basin-forming events, the deeper fill being found in the younger basins. Nearly thirteen hundred separate eruption locations have been identified. This number does not include those buried by their own or younger erupted materials. The true number of eruption sites may be closer to thirty thousand.

Returned lunar samples have radiometric ages that fall within a range of 3 to 4 billion years. The presence of basaltic fragments in breccias produced by the large-basin-producing impacts dating from about 4 billion years, along with dark materials excavated by impacts from beneath a younger impact depositional cover, indicates that the age of volcanism extends further back in time than the ages of the returned samples. Furthermore, ages determined by crater-counting techniques have indicated the presence of lavas perhaps as young as 2 billion years. Thus, although now long absent, the majority of volcanic flows on the Moon took place over a time span of about 1 billion years. Additionally, crater-counting data indicate that individual maria may have witnessed eruptions for a similar time span, the older materials being deeply buried near the basins' centers. Crater counting and superposition relationships have led to the construction of a relative timescale based on the time of formation of large impact craters and basins. Subdivisions of this timescale in order of decreasing age are pre-Nectarian, Nectarian, Early Imbrian, Late Imbrian, Eratosthenian, and Copernican. Two-thirds of the nearside surface of the maria is of Late Imbrian age; much of the remainder is of Eratosthenian age. Early Imbrian, Nectarian, and pre-Nectarian lavas could have been buried by impact ejecta and younger lavas; Copernican-aged lavas are restricted to small areas of the western near side.

Morphological evidence for volcanism is found in the form of lava-flow fronts, sinuous rilles, mare domes, cones, and pyroclastic deposits. Measured flow fronts have heights which average about 30 meters but range from 10 to 63 meters. It is highly probable that thinner flows once existed, but they were subject to obliteration by more than 2 billion years of erosion from meteorite impacts. Flow fronts outline individual eruptions more than several hundred kilometers in length. The great size of these features is attributable to the extremely low viscosities of erupted materials and their high-volume output rate. Sinuous rilles superficially resemble terrestrial river channels and range from hundreds of meters to 3 kilometers in width and a few kilometers to 300 kilometers long. These features are believed to be lava channels or collapsed lava tubes. The Apollo 15 mission included exploration at the edge of one of these structures, Hadley Rille. More than three hundred mare domes have been identified on the Moon. They

The volcanic plains of the Moon, called "seas" or (Latin) "maria," can be seen in this image from Apollo 15. These smooth surface areas form nearly one-fifth of the surface. (NASA)

Lunar Maria

Name	Latitude	Longitude	Diameter	English Name
Mare Anguis	22.6N	67.7E	150.0	Serpent Sea
Mare Australe	38.9S	93.0E	603.0	Southern Sea
Mare Cognitum	10.0S	23.1W	376.0	Sea That Has Become Known
Mare Crisium	17.0N	59.1E	418.0	Sea of Crises
Mare Fecunditatis	7.8S	51.3E	909.0	Sea of Fecundity
Mare Frigoris	56.0N	1.4E	1,596.0	Sea of Cold
Mare Humboldtianum	56.8N	81.5E	273.0	(Alexander von Humboldt, German natural historian, 1769-1859)
Mare Humorum	24.4S	38.6W	389.0	Sea of Moisture
Mare Imbrium	32.8N	15.6W	1,123.0	Sea of Showers
Mare Ingenii	33.7S	163.5E	318.0	Sea of Cleverness
Mare Insularum	7.5N	30.9W	513.0	Sea of Islands
Mare Marginis	13.3N	86.1E	420.0	Sea of the Edge
Mare Moscoviense	27.3N	147.9E	277.0	Sea of Muscovy
Mare Nectaris	15.2S	35.5E	333.0	Sea of Nectar
Mare Nubium	21.3S	16.6W	715.0	Sea of Clouds
Mare Orientale	19.4S	92.8W	327.0	Eastern Sea
Mare Serenitatis	28.0N	17.5E	707.0	Sea of Serenity
Mare Smythii	1.3N	87.5E	373.0	(William Henry Smyth, British astronomer, 1788-1865)
Mare Spumans	1.1N	65.1E	139.0	Foaming Sea
Mare Tranquillitatis	8.5N	31.4E	873.0	Sea of Tranquillity
Mare Undarum	6.8N	68.4E	243.0	Sea of Waves
Mare Vaporum	13.3N	3.6E	245.0	Sea of Vapors

Source: National Space Science Data Center, NASA Goddard Space Flight Center.

have shapes and dimensions comparable to small terrestrial shield volcanoes.

Conical structures are common in the Marius Hills volcanic complex within Oceanus Procellarum but are relatively rare elsewhere. It is believed that these structures are the lunar equivalent of a terrestrial Strombolian eruption style. The Marius Hills area had been seriously considered as a landing site for Apollo astronauts, but it did not make the cut when three Apollo missions were canceled as a result of budget cuts.

Pyroclastic deposits on the Moon can be subdivided into two major groups: dark halo deposits and regional dark mantle deposits. Dark halo deposits extend to ranges of about 5 kilometers from an endogenic crater. Dark mantle deposits cover areas of up to 40,000 square kilometers. The Apollo 17 mission included the only geologist astronaut, Harrison "Jack" Schmitt, to fly to the Moon to do fieldwork on its surface. On that final lunar landing mission, the astronauts returned samples of this pyroclastic material in the form of orange "soil" found in the Taurus-Littrow valley. Stratigraphy and compositional variations of lunar pyroclastic materials suggest an age range comparable to the mare basalts.

Compositionally and texturally, lunar mare samples returned to Earth are basaltic lavas and glasses. Samples from each landing site are unique in terms of their major and minor element chemistry. Additionally, different chemistries can be recognized in the basalts at each site. General distinctions have been drawn between high, intermediate, low, and very low titanium basalts and feldspathic basalts. It must be remembered that samples have been returned from only six mare sites. Fortunately, basalts are rich in the transition group metals, particularly iron and titanium. It has been possible to map the distribution of compositionally different basaltic materials across the nearside of the Moon on the basis of their spectral characteristics. More than one dozen spectrally different units have been identified. There appear to be no simple age-composition relationships or composition-location relationships. For example, there are both old and young titanium-rich basalts, and titanium-rich basalts in both the eastern and the western hemispheres of the Moon.

There are also dome-shaped features on the lunar surface whose shapes and spectral characteristics serve to distinguish them from mare domes. These features have heights of up to slightly more than 1 kilometer and areas of up to 500 square kilometers. They are all close to the highland-mare boundary in Oceanus Procellarum. Their spectra have suggested a comparison with KREEP basalts, which are rich in potassium, rare Earth elements, and phosphorus. If of volcanic origin, their shapes would indicate formation by higher-viscosity materials or at lower volumetric eruption rates than mare basalts. Stratigraphic analysis indicates that they were emplaced at the lunar surface at the same time that basaltic eruptions were taking place elsewhere.

In summary, volcanic activity, perhaps dominated by KREEP-rich materials, preceded development of younger large impact basins. Cavities created by the large impacts became the sites of basalt deposition, and deeper, central portions of the basins were the first to be filled. Widespread, flood-type volcanism was locally accompanied by the eruption of more volatile-rich pyroclastic material. With progressive infilling, the magma had an increasingly difficult task in penetrating the thick, high-density basin fill,

so volcanism shifted to the periphery of the basins, found outlets in the floors of circum-basinal craters, and overflowed into the surrounding terrains.

Methods of Study

Locations and stratigraphic relations of the lunar maria have been determined by geologic mapping using Earth-based and orbital photographs as the database. Relative ages of different surface units have been determined by measuring the number of impact craters of a given size within a specified area or by determining the extent to which craters of a given size have been modified with respect to fresh, similarly sized craters. Both of these techniques rely on the fact that the Moon is under constant bombardment by meteorite particles, so that the longer an area has been exposed at the lunar surface, the more craters it will have and the more the pre-impact surface will be affected by erosion and deposition. In many cases it is possible to identify secondary craters created by material excavated by a primary impact. These secondary craters (and the primary crater) have to be younger than the material into which they impacted. By mapping these relationships, scientists have established the lunar stratigraphic column.

The advantage of the return of lunar samples was that the relative ages calculated by the cratering techniques could be related to the absolute ages determined by radiometric dating. Furthermore, exposure times (rather than crystallization or metamorphic ages) of the returned samples at the lunar surface have been established by determining their amounts of solar-derived particles. Sizes of such structures as lava-flow fronts have been determined by measuring the lengths of shadows and through the use of laser altimeter carried aboard orbiting spacecraft. These data have been compiled to make topographic maps.

Many techniques have been employed to determine the thickness of the mare basalts. One geophysical technique relies on the fact that the mare basalts have a greater density than the surrounding highlands materials. This greater density results in the maria exerting a greater gravitational pull on a spacecraft, causing minute changes in its orbital motions. This phenomenon led to the discovery of strong pulls over the younger circular impact basins and the formulation of mascons (short for mass concentrations). Morphometric techniques for determining mare thicknesses rely on the fact that impact craters have regular and predictable dimensional characteristics. For example, a fresh, bowl-shaped crater has a depth equal to one-sixth its diameter. If an impact crater has penetrated a mare surface and excavated pre-mare materials from beneath, then the mare must be less thick than one-sixth of the crater diameter. Similarly, if all pre-mare craters with less than a specific diameter have been buried by lava, then, again, a minimum depth can be established. At the larger end of the scale, it is possible to take a topographic map of a relatively unflooded young basin, such as the Orientale Basin, and artificially raise the lava level parallel to the contours until the basin looks like the more deeply flooded basins. Height difference is then a measure of the mare infill thickness in the more deeply flooded basin.

Composition of mare basalts has been established by standard geochemical techniques applied to returned samples. More Moon-wide data have been obtained from orbital geochemical experiments. Information on radon and polonium variations was gathered by alpha-particle spectrometry; uranium, potassium, and thorium concentrations, and the elemental abundances of oxygen, silicon, iron, magnesium, and titanium were determined by gamma-ray spectrometry; and aluminum, silicon, and magnesium variations were determined using X-ray fluorescence data. Various photographic and reflectance spectroscopy techniques have been employed by Earth-based observers to determine transition element variations, mineral compositions, and glass contents of the mare surfaces facing the Earth.

Lunar studies ceased after the Soviet Luna 24 mission robotically returned to Earth samples from the Sea of Crises (Mare Crisium) in 1976. The Moon was ignored for nearly two decades in terms of planetary science programs involving robotic spacecraft. Then in the 1990's the Ballistic Defense Organization sent the Clementine spacecraft to test new sensors and in the process map lunar resources. NASA followed with the more capable Lunar Prospector. Clementine and Lunar Prospector returned tantalizing suggestions of water ice on the surface of the Moon in permanently shadowed areas. Data from the Clementine spacecraft in 1994 appear to indicate an ice field near the Moon's south pole. Five to ten meters deep and sixteen thousand square kilometers in area, the ice field is mixed with soil; it could be a valuable resource for crewed lunar bases. That finding was substantiated by NASA's Lunar Prospector spacecraft, which orbited the Moon from January, 1998, to June, 1999, and determined the chemical composition of the lunar surface using alpha particle, neutron, and gamma-ray spectrometers. Renewed interest in the Moon followed. The Chinese, Indian, European, and Japanese space programs sent spacecraft into orbit about the Moon in the first decade of the twenty-first century, and NASA prepared to launch the Lunar Reconnaissance Orbiter in 2009.

In the aftermath of the *Columbia* accident (2003), NASA was directed by the Bush administration to fulfill its Space Vision for Exploration, which calls for a return to the Moon during which astronauts would begin setting up a lunar base. Although the plan, the Constellation program, calls for the base to be established near the Moon's south pole, the infrastructure is to be designed so that exploration of the surface far from the base is possible; this could include sorties to the lunar maria. In any event, with China planning to send its taikonauts (the Chinese equivalent of astronauts) to the Moon as well, activity in orbit and on the surface of the Moon should increase dramatically, increasing our understanding of the Moon's history and geology.

Context

Maria form the least rugged terrain on the Moon, so they are conducive to the safe landing of spacecraft. These flat areas can also facilitate lunar exploration through the use of surface craft. They are most abundant on the Earth-facing hemisphere, where continuous telecommunications are possible with the Earth. Excavation of the maria to produce dwellings would not be very difficult. Alternatively, natural shields to solar radiation, such as lava tubes, could be turned into habitation sites. Several techniques have been proposed for the mining of mare basalts to be used as raw materials for the construction of spacecraft in Earth orbit and to provide sustenance for lunar inhabitants. These factors make the maria primary candidates for the establishment of a permanent, crewed lunar base.

A farside mare site would be an ideal location for the construction of an astronomical observatory. Communications with Earth-based stations would necessitate a more complex satellite system than that required by a nearside base, but at the same time, the shielding from terrestrial electromagnetic radiation provided by the bulk of the Moon would enable clear views of distant galaxies.

From a scientific viewpoint, the surface materials of the maria provide a diary of solar activity and small meteorite impacts extending over the past few billion years. On Earth, by contrast, this information is less available because it has been removed from rocks by erosion, deposition, and the recycling of oceanic plates. Detailed study of the maria can therefore furnish information about what was taking place in the near-Earth solar system from the time of emplacement of the oldest surviving terrestrial rocks through the development of unicellular organisms to the arrival of humans. These data cannot be obtained from the surface of Venus, because of the thick atmosphere of that planet; nor will Mars serve, because of the effects of wind erosion, ice formation, and, perhaps, past erosion by river systems. Therefore, the Moon is a unique natural laboratory, and the maria provide favorable sites for a laboratory inhabited by humans. With the declaration of the Space Vision for Exploration, some of these grand plans may come to fruition during the twenty-first century.

James L. Whitford-Stark

Further Reading

Baldwin, Ralph B. *The Measure of the Moon*. Chicago: University of Chicago Press, 1963. A classic book used by those directly involved with the scientific planning of the crewed and uncrewed lunar landings. Surprisingly much of this tour-de-force study of the Moon remains valid. Most valuable for the correlations it draws between impact-cratering phenomena and human-made explosions. Suitable for advanced high school and college students.

Beattie, Donald A. *Taking Science to the Moon: Lunar Experiments and the Apollo Program*. Baltimore: Johns Hopkins University Press, 2003. Explains the science gleaned from the Apollo lunar landings, including the Apollo Lunar Surface Science Experiment Packages (ALSEPs) and their results.

Lunar and Planetary Institute, Houston, Texas. *Basaltic Volcanism on the Terrestrial Planets*. Elmsford, N.Y.: Pergamon Press, 1981. An indispensable text for any study of planetary geology. Entire chapters are devoted to petrology and geochemistry, remote sensing, surface morphologies, radiometric dating, and crater studies. For college-level readers.

Mutch, Thomas A. *Geology of the Moon*. Rev. ed. Princeton, N.J.: Princeton University Press, 1972. A readable textbook for advanced undergraduate and graduate students. Describes the Moon from a stratigraphic viewpoint, with the Apollo results forming the last chapter.

Schmitt, Harrison J. *Return to the Moon: Exploration, Enterprise, and Energy in the Human Settlement of Space*. New York: Copernicus Books, 2006. A scientific and economic plan for prolonged exploration and exploitation of lunar resources, written by the only geologist to land on the Moon and perform field geology and collect documented samples for return to Earth. Discusses the use of helium-3 mining to supply future energy production systems on Earth and thereby finance and systematically expand lunar settlements and research posts.

Schrunk, David, et al. *The Moon: Resources, Future Development, and Settlement*. New York: Springer Praxis, 2007. Examines what is needed to develop human settlements on the lunar surface, beginning with renewed robotic exploration. Explains the benefits of a robust spacefaring civilization.

Schultz, Peter H. *Moon Morphology: Interpretations Based on Lunar Orbiter Photography*. Austin: University of Texas Press, 1974. A post-Apollo photographic encyclopedia of lunar surface features. Aimed primarily at a research-level audience.

Spudis, Paul D. *The Geology of Multi-ring Impact Basins: The Moon and Other Planets*. New York: Cambridge University Press, 1993. A comprehensive geological study of large impact craters formed on the planets and their satellites. Well illustrated with photographs and diagrams. Suitable for both the lay reader and scientific audiences.

Tumlinson, Rick N., and Erin Medlicott, eds. *Return to the Moon*. New York: Collector's Guide Publishing, 2005. A series of essays by lunar experts that examine the Space Vision for Exploration. A broad examination of a return to the Moon, this time to stay and set up lasting infrastructure.

Wilhelms, Don E. *The Geologic History of the Moon*. Professional Paper 1348. Denver, Colo.: U.S. Geological Survey, Books and Open-File Reports Section, 1987. This volume synthesizes the author's vast accumulated knowledge of mapping the lunar surface with information derived from every other field of lunar research. An essential source for budding lunar scientists and a reference source for specialists needing to put their work in perspective. Heavily illustrated.

_____. *To a Rocky Moon: A Geologist's History of Lunar Exploration*. Phoenix: University of Arizona Press, 1994. Numerous books portray the Apollo astronauts, who successfully completed six lunar landings, with heroic prose. Others describe the science learned from Apollo investigations and returned Moon rocks and soil samples. This book is a unique geology-driven explanation for the way individual missions were planned and carried out for maximum science return. Includes descriptions of real-time evaluations of field geology aspects on each Apollo mission.

LUNAR REGOLITH SAMPLES

Category: Natural Planetary Satellites

Initial study of lunar soils focused on ensuring the safety of the crewed Apollo spacecraft upon landing. Returned soil samples have since been analyzed to determine the origin and evolution of the Moon. Future studies will assess the suitability of lunar resources for utilization in construction of a Moon base and as raw materials for space manufacturing.

Overview

Existence of a "soil," or layer of small particles covering the lunar surface, was inferred prior to the first spacecraft's landing on the Moon. At full Moon, the lunar surface is observed to be bright from edge to edge, exhibiting only minimal "limb darkening" (a decrease in intensity of the light reflected near the edges of a smooth sphere). This led early observers to conclude that the uppermost surface layer of the Moon was porous on the centimeter scale, suggesting a surface dust layer. Determination of the thickness and physical properties of this dust layer was important to the success of the crewed lunar landings. Cornell astronomer Thomas Gold had postulated that the Apollo Lunar Module might sink in a thick surface dust layer that could not bear the weight of the vehicle.

Early in situ investigations of the properties of the lunar soils were conducted by lunar soft-landing spacecraft in the Soviet Luna series and the Surveyor spacecraft series launched by the National Aeronautics and Space Administration (NASA). On February 3, 1966, Luna 9, the first spacecraft to soft-land on the Moon, returned panoramic photographs of the surface and demonstrated that the soil was firm enough to support the 100-kilogram payload without noticeable effect. Surveyor spacecraft and the Luna 13, 17, and 21 soft-landers carried instruments that determined the lunar soil's composition and physical properties.

Nevertheless, the return of lunar samples to Earth for laboratory analysis offered significant advantages; earthbound instruments were not limited, as were the lightweight ones suitable for space flight. The first samples of lunar soil were returned to Earth by Apollo 11. During their stay on the lunar surface, the Apollo 11 astronauts collected 22 kilograms of lunar material. Of this sample, 11 kilograms were categorized as "fines" (particles smaller than 1 centimeter).

In addition to surface soil samples, Apollo astronauts obtained "core samples," cylindrical samples of lunar soil taken by pushing a tube vertically into the lunar surface. Preserved layers in cores provide information on the rates of depositing and mixing of the soil. Apollo 11 astronaut Edwin E. Aldrin collected two core samples in 2-centimeter-diameter tubes pushed down into the lunar surface. The first core, about 10 centimeters long, contained 51 grams of material, while the second, measuring 13.5 centimeters, weighed 65 grams.

Apollo 11 lunar samples were returned to Earth on July 24, 1969, and flown to the Manned Spacecraft Center (MSC, later renamed Johnson Space Center), in Houston, Texas. There the samples were placed in quarantine in the Lunar Receiving Laboratory (LRL) for a period of one month while biological analyses were conducted. The Lunar Sample Preliminary Analysis Team (LSPAT), consisting of MSC scientists and visiting scientists, was permitted to study samples under controlled conditions during the quarantine period. They first exposed small chips from the samples to nitrogen, oxygen, and air at various humidities to ensure that laboratory analysis conditions did not cause adverse reactions or sample deterioration. Within the LRL, samples were characterized by mineralogical and chemical techniques. In addition, experiments were performed to determine the effects of cosmic-ray exposure on the lunar material, the organic carbon content, and the noble gas concentrations. Additional experiments searched for "magnetic monopoles" (particles of isolated magnetic charge whose existence is postulated by elementary particle physicists).

Following quarantine, lunar samples were made available to about 110 scientists, selected by the Office of Space Science and Applications, to perform a variety of experiments. These scientists—from twenty-one universities, two industrial facilities, three private institutions, and ten government laboratories—included twenty-seven scientists from England, Germany, Canada, Japan, Finland, and Switzerland. Because of the fineness of the returned material, the scientists developed new techniques and instruments to perform a variety of experiments on the lunar samples. After six months, they met to discuss their results at the Apollo 11 Lunar Science Conference, held at MSC from January 5 to January 8, 1970. The Apollo 11 Lunar Science Conference evolved into an annual meeting at which scientists from around the world report their latest results on lunar sample research and planetary science. NASA continues to allocate samples collected

Apollo 11 delivered the first human beings to the surface of the Moon in 1969 and initiated a series of missions, several of which returned regolith samples. Here astronaut Buzz Aldrin salutes an American flag, left on the surface. The flag had to be backed by a structure that would make it appear to be waving in a nonexistent lunar wind. (NASA)

during the Apollo missions as new instruments or new techniques warrant further experiments.

Fines were shown to consist of a mixture of glassy materials and small crystal fragments. Within the cores, the majority of the particles ranged in size from 1 millimeter down to 30 micrometers. Glasses exhibited a variety of colors from pale or colorless to gray to wine red, orange, green, brown, and yellow. Crystal fragments were dominated by the minerals plagioclase, clinopyroxene, ilmenite, and olivine.

Lunar material from five additional sites on the Moon was returned to Earth by the Apollo 12, 14, 15, 16, and 17 missions. Core samples up to 2.6 meters in length were obtained. More than 380 kilograms of lunar material were returned to Earth by the Apollo program, more than half of which was collected on the final two Apollo flights.

The Soviet Union's recovery of lunar materials involved the uncrewed spacecraft in the Luna series. Luna 16, which landed on the Moon on September 20, 1970, returned a single 35-centimeter drill core containing

101 grams of soil on September 24. Luna 20 returned a similar sample in February of 1972. The more advanced Luna 24 spacecraft, which landed in the Sea of Crises area of the Moon on August 18, 1976, returned a two-meter core sample.

These lunar samples have been made available to scientific investigators throughout the world. An agreement between the National Aeronautics and Space Administration (NASA) and the Soviet Academy of Sciences in 1971 provided for the exchange of Apollo and Luna samples, allowing all investigators to have access to the soils collected at the nine lunar sites sampled by either Apollo or Luna missions.

Proposals to establish a lunar base and to develop space manufacturing facilities have focused attention on the lunar soils as construction material and as raw material for the manufacturing process. Procedures for the extraction of aluminum from the plagioclase, titanium from the ilmenite, and magnesium from the olivine grains in the soils have been described. As a by-product of these extractions, silicon and oxygen can also be produced. Although lunar soils contain essentially no water, hydrogen implanted by the solar wind might be obtained from the ilmenite and then be used to react with oxygen to produce water. Use of implanted hydrogen as a rocket fuel has also been proposed. Thus, it appears that the high cost of launching materials from Earth's surface for space manufacturing can be circumvented by acquiring the bulk of the raw materials in the low-gravity environment of the Moon.

Knowledge Gained

Lunar regolith (a blanket of broken fragments ranging from dust- to meter-sized blocks of rock) was found to cover almost all the Moon to an estimated depth of a few meters. The fine component of the regolith, generally referred to as soil, was shown to be very different from terrestrial soil. On Earth, the soil is formed by the complex action of atmospheric weathering and biological activity. Lunar soil is produced by bombardment of the surface rocks by meteorites and micrometeorites, producing craters and microcraters; the impact debris is then scattered over the lunar surface.

Chemical composition of the soil was discovered to differ slightly from that of the lunar rocks. By subtraction of the rock composition from that of the soil, the chemical composition of the added component was found to be similar to the composition of primitive stone meteorites that fall on Earth. Thus, meteoritic

bombardment was proved to be an important alteration process on the Moon.

Analyses of the layered structure of the tube samples suggested that large meteorite impacts have thrown layers of soil across the lunar landscape. The top surface of this layer has been stirred by impacts of more numerous micrometeorites. The layer structure observed in long core tubes collected on later Apollo missions agrees well with computer simulations of the rate of soil mixing, calculated assuming the rate at which small particles strike the Moon has been approximately equal to the measured current rate for millions of years. Radiation damage in grains of the lunar soil, as well as in lunar rocks, suggests that the rate of emission of heavy ions by the Sun has also been relatively constant over the past few million years.

Layers in the lunar soil record the history of bombardment of the Moon, with a single core providing samples of material ejected from craters far apart on the lunar surface. One Apollo 12 core contains a layer, light gray in color and rich in silica, that has been identified as ejecta (material thrown out by volcanic eruptions) from the crater Copernicus, 75 kilometers from the core site. Thus, Apollo core samples have provided information on the composition and mineralogy of locations far from the sampling site.

Context

Both the Apollo and Luna programs returned samples of the lunar suface material. Though these lunar rocks could be compared to similar terrestrial rocks, lunar soil was clearly different from its terrestrial counterpart. Analysis of soils and the stratigraphy of core samples provided information necessary to confirm the importance of meteoritic bombardment in producing rock fragments which make up lunar soil and in stirring the upper layer of the soil once it is placed on the lunar surface. Measurement of effects of cosmic rays in an Apollo 15 core sample indicated that layered structures within the lunar soil can be preserved undisturbed on the lunar surface for periods of 500 million years. Cores established that the lunar erosion process proceeds at a rate of between 1 and 2 millimeters per million years, about one thousand times slower than on Earth.

The depth of the lunar soil layer confirmed that the rate at which meteorites and micrometeorites are hitting the Moon has remained essentially unchanged over the past four billion years. The presence of ion damage caused by solar flares in soils buried for millions of years also confirmed that the flux of charge particles from the Sun has been relatively constant.

The chemical composition of the bombarding material was also deduced from the difference between soil and bulk rock compositions. Several distinct compositions were observed, but all were generally similar to the known meteorites collected on Earth.

Studies of returned lunar soils also demonstrated their potential as a raw material for lunar base construction and for space manufacturing. Proposals for lunar mining have generally assumed the soils as the starting material. Since the delivery of large masses of material from the lunar surface to near-Earth orbit requires less energy than delivery of the same mass of material from the surface of Earth, lunar soils are expected to play a significant role in the industrialization of space. The Apollo and Luna soil samples are available on Earth to test and perfect extraction techniques in Earth laboratories so that equipment appropriate for the first lunar mining facility can be designed.

No soil samples were returned to Earth after the Soviet Luna 24 mission in 1976, but in the 1990's a pair of spacecraft, Clementine and Lunar Prospector, mapped the surface for mineral content. A renewed interest in the Moon led the European Space Agency (ESA), Japanese Aerospace Exploration Agency (JAXA), the Chinese National Space Administration (CNSA), and the Indian Space Research Organization (ISRO) to send probes to the Moon in the first decade of the twenty-first century. These spacecraft were named SMART 1 (ESA), Kaguya (JAXA), Chang'e 1 (CNSA), and Chandrayaan 1 (ISRO).

In 2009, NASA was preparing to launch the Lunar Reconnaissance Orbiter to begin a comprehensive mapping of lunar resources and assist in the determination of the best location near the Moon's south pole for the construction of a permanently occupied lunar base. That goal was the first objective of the Bush administration's Vision for Space Exploration set up in the aftermath of the 2003 *Columbia* accident. NASA was charged with a return of astronauts to the Moon by 2019, the fiftieth anniversary of the Apollo 11 lunar landing. However, the Chinese also stated a firm commitment to evolve their early Shenzhou piloted missions into eventual flights to the Moon, and there is a very real chance that when American astronauts return to the Moon, Chinese taikonauts (the Chinese equivalent of astronauts) will already have visited the lunar surface. Regardless of which nations return to the Moon, the renewed interest in lunar exploration portends advances in understanding of our solar system and a first step in lunar colonization.

George J. Flynn

Further Reading

Baldwin, R. A. *The Measure of the Moon*. Chicago: University of Chicago Press, 1963. The first nine chapters outline the characteristics of both lunar and terrestrial, human-made, and natural craters. Written in pre-metric-system times, this classic text requires mathematical (though simple) conversions. Suitable for advanced high school and college students.

Beattie, Donald A. *Taking Science to the Moon: Lunar Experiments and the Apollo Program*. Baltimore: Johns Hopkins University Press, 2003. Explains the science gleaned from the Apollo lunar landings, including the Apollo Lunar Surface Science Experiment Packages (ALSEPs) and their results.

French, Bevan M. *The Moon Book*. New York: Penguin Books, 1977. A classic work that describes the origin of the Earth-Moon system and evaluates these explanations in the light of the lunar-sample results from the Apollo and Luna missions.

Hartmann, William K. *Moons and Planets*. 5th ed. Belmont, Calif.: Thomson Brooks/Cole, 2005. An updated version of a classic text that covers all aspects of planetary science. Particularly strong in its presentation of Earth-Moon science. Uses the comparative planet approach, which means there are sections on each of the planets in chapters such as "Planetary Interiors," "Planetary Surfaces," and "Planetary Atmospheres."

Jefferis, David. *Return to the Moon*. New York: Crabtree Children's Books, 2007. An explanation of contemporary plans to return American astronauts to the Moon as part of Constellation program operations with the Orion Crew Exploration Vehicle and Altair landing craft. Aimed at a younger audience.

King, Elbert A., Jr. *Space Geology: An Introduction*. New York: John Wiley and Sons, 1976. This 349-page text, intended for advanced students and written by the curator of NASA's Apollo 11 Lunar Receiving Laboratory, includes a well-illustrated section describing the analyses of lunar soils.

Levinson, Alfred Abraham, ed. *Apollo 11 Lunar Science Conference: Proceedings*. 3 vols. Elmsford, N.Y.: Pergamon Press, 1970. This 2,490-page work

includes articles by the Apollo 11 principal investigators that report the techniques employed and the
preliminary results obtained in their analyses of the
Apollo 11 lunar samples.

Mason, Brian, and William G. Melson. *The Lunar
Rocks*. New York: Wiley-Interscience, 1970. Written
shortly after the Apollo 12 sample return, this book
analyzes the results reported on the Apollo 11 and
12 samples. Includes a chapter describing the fines
(lunar soils).

Schmitt, Harrison J. *Return to the Moon: Exploration,
Enterprise, and Energy in the Human Settlement of
Space*. New York: Copernicus Books, 2006. A scientific and economic plan for prolonged exploration and exploitation of lunar resources written by
the only geologist to land on the Moon. Talks about
helium-3 mining, aimed at supplying future energy
production systems on Earth and thereby financing
and systematically expanding lunar settlements and
research posts.

Taylor, S. R. *Planetary Science: A Lunar Perspective*.
Houston: Lunar and Planetary Institute, 1982. This
well-illustrated college-level work, written by a
member of the Lunar Sample Preliminary Analysis
Team for Apollo 11, describes the geological processes active on moons and planets in the light of
the discoveries from the Apollo lunar samples.

Wilhelms, Don E. *To a Rocky Moon: A Geologist's History of Lunar Exploration*. Phoenix: University of
Arizona Press, 1994. Numerous books portray the
Apollo astronauts, who successfully completed six
lunar landings, with heroic prose. Others describe
the science learned from Apollo investigations and
returned Moon rocks and soil samples. This book
is a rather unique personal account in that it is a
geology-driven explanation for the way individual
missions were planned and carried out for maximum
science return. Includes descriptions of real-time
evaluations of field geology aspects on each Apollo
mission.

Wood, John A. "The Lunar Soil." *Scientific American*
223 (August, 1970): 14-23. In this article, one of the
principal investigators of the Apollo lunar samples
describes the results of the first analyses of the returned lunar soil samples. Well illustrated and intended for the general audience.

LUNAR ROCKS

Category: Natural Planetary Satellites

*Lunar rocks are among the materials brought to Earth
from the surface of the Moon during the Soviet robotic
Luna and the American Apollo crewed Moon missions of
the 1960's and 1970's. These samples were analyzed by
international teams of scientists in an effort to enhance
understanding of the physical nature of the Moon and
its origins. All samples were cataloged according to criteria relating to mineralogy, crystallography, geochro-
nology, geochemistry, magnetism, radioactivity, and
other characteristics.*

Overview

When direct exploration of the surface of the Moon began
during the 1960's, scientists around the world looked forward with great anticipation to the study and analysis of
materials brought back to Earth from the lunar surface
by the crewed space missions of the Apollo program and
the uncrewed programs of the Soviet Union such as the
Luna missions. As new and different material arrived at
the completion of each successive mission, an astonishing
wealth of information about the origin and physical character of the Moon began to accumulate. A great deal of the
Apollo lunar samples remain to be studied even after forty
years or more. A conscious effort had been made to save
large portions of the Moon rock collection for new technologies that could be employed to study them in ways
not possible back in 1969 through 1972, the era of the
Apollo lunar landings.

Using the latest scientific techniques to analyze soil
and rock gathered during the missions, researchers were
able to identify the basic elements and minerals that
constitute the Moon's surface. Using that knowledge in
combination with other information gathered by remote-
sensing technology and sophisticated photography techniques over several decades, they were able to extrapolate information about the origin of the Moon. Perhaps
most important, they were able to determine with some
accuracy a chronology of events in lunar evolution. While
there remain many unsolved mysteries about the Moon
and its history, it is remarkable how much information
was gathered from the study of so-called Moon rocks.

"Moon rocks" are defined as surface materials that exceed 1 centimeter in diameter. Anything smaller is considered lunar "soil," or regolith, although lunar regolith is
unlike Earth soil, which usually contains large amounts of

decayed organic material and moisture. On the Moon, the lack of atmosphere and organic material gives the soil a composition similar to dry, clean sand. It is probably powdered rock, created when large objects such as meteoroids impacted violently with the surface of the Moon over many millennia.

Rocks on the surface of the Moon are rich in calcium and aluminum. Below 20 kilometers, the surface is not broken as extensively as it is near the top. The upper mantle is from 200 to 300 kilometers thick and contains high concentrations of magnesium and iron silicate, pyroxene, and olivine. There is reason to believe that the core is iron-rich and produced a global magnetic field during the early days of the Moon's history. Most of the radioactive materials on the Moon are located at or near the surface. Two major activities that have resulted in the character of the lunar surface are volcanism and impact events. Lowland regions exhibit evidence that there has been great volcanic activity on the Moon, and across the lunar landscape there is evidence that impacts have resulted in broad redistribution of rock and soil. Occasionally that activity has resulted in the breaking up of bedrock and basalt breccias that formed prior to later redistribution, recrystallization, or both.

Each of the lunar missions that returned to Earth with samples of Moon rock expanded the body of knowledge about the geological character of the Moon's surface. Each mission landed at a different location on the lunar surface to ensure that as broad a collection of materials as possible would be acquired given the limited number of

Astronaut Harrison Schmitt stands at the foot of an outcrop of boulders near the Taurus-Littrow landing site of Apollo 17. (NASA/E. Cernan)

missions planned and the limited payload capacity available on returning spacecraft. For the purposes of lunar exploration, two primary regions of the Moon were identified. Generally speaking, the lunar highlands are regions that appear light on the surface on the Moon. The maria constitute that portion of the lunar surface that is dark. The highlands are generally composed of remnants of ancient volcanic cones and the ridges of giant impact craters. The maria are surface areas characterized by volcanic chamber depressions and impact crater basins. Analysis of rocks collected at various spots within each of the two regions demonstrates that each has unique physical characteristics.

Moon rocks are of two main types, crystallized (or igneous) and breccias. The igneous rock encountered on the lunar surface is most often volcanic in origin and appears to have been scattered over wide areas during violent eruptions. Some specimens are believed to have been thrown between 100 and 1,000 kilometers, either during

eruptions or as a result of meteor impacts. Certain highland samples contain evidence of the existence of plutonic rock, formed when volcanic material cools below the surface.

The Apollo 17 crew, working with the benefit of knowledge gained in previous crewed and uncrewed missions, further subcategorized lunar rock as follows: basalts; dark matrix breccias; glass-bonded agglutinates; vesicular green-gray breccias; blue-gray breccias; layered, foliated, light-gray breccias; and brecciated gabbroic rocks. The distribution of Moon-rock material is different for highland and mare regions. In the highlands, the surface consists of approximately 10 percent plutonic rock, 85 percent breccias, and 5 percent volcanic material, primarily basalts. Mare regions contain 90 percent volcanic basalts and 10 percent breccias.

It is important to note that the number of minerals identified on the Moon is dwarfed by the number found on Earth; this difference probably results from the Moon's lack of atmosphere, particularly oxygen and moisture, which often interact with elements in the weathering process to form minerals. There are some two thousand minerals on Earth, whereas there are only two hundred known to exist on the Moon. The number of primary minerals, however, is similar on both bodies. The principal minerals discovered on the Moon at the regolith level include clinopyroxene, plagioclase, olivine, ilmenite (by far the most abundant, 15-20 percent by volume), trydymite, cristobalite, and orthopyroxenes. Several new minerals have been identified, including armalcolite, pyroxferroite, and tranquillityite. It is interesting to note that armalcolite was named for the three astronauts on the Apollo mission that returned with it: Neil Armstrong, "Buzz" Aldrin, and Michael Collins. Armalcolite has since been found on Earth; it is formed when crystallization occurs in the absence of moisture and oxygen.

Ancient orange glass beads, thought to have been spewed out during volcanic eruptions, have given scientists a glimpse of what constitutes the core of

During the Apollo 16 mission, Commander John W. Young took the Lunar Roving Vehicle for a spin near the Descartes landing site; 90 percent of soil and rock samples brought from this region had high concentrations of aluminum. (NASA/Charles M. Duke, Jr.)

the Moon's interior. These beads contain high levels of lead, zinc, tellurium, and sulfur and are thought to have originated as deep as 300 kilometers below the surface of the Moon. Lunar rock samples have been individually but not uniformly rich in such elements as aluminum, magnesium, potassium, phosphorus, titanium, iron, chromium, and zirconium. Some lunar rocks are referred to as glassy agglutinates. These rocks are believed to have been formed when volcanic action or meteoric impact caused fine rock dust to weld together, a process known as impact melting. Their appearance ranges from dark opaque to transparent. Other rocks have different origins but are covered by glassy materials thought to have formed and been distributed in the same manner. While the great majority of collected lunar rocks are igneous, there is some evidence that a metamorphosis has occurred as a result of shock settling of soil, caused by meteor impact and vibrations that occur during the course of violent volcanic activity.

Breccias appear to contain material that has been ejected by impacts or volcanic eruption. There are two types: soil breccia and ejecta blanket breccia. Soil breccia is composed of the same materials that exist as soil in a particular region and probably results from the shock of meteor impact. Soil breccia is also characterized by weak cohesion. Ejecta blanket breccias, by contrast, are thought to have been formed when volcanically ejected materials fused together to form a layer of rock that later underwent various forms of thermal metamorphism. The composition of such rock varies with the depth at which it has been found or at which it is thought to have originated.

The primary source of lunar erosion appears to have been impact cratering, although solar wind and cosmic-ray effects have also altered the character and composition of surface materials. Some lunar samples show cosmic-ray bombardment over long periods of time that has resulted in the alteration of the surface of the sample. The same activity has been useful in helping scientists determine how long a particular sample fragment was in the same location and at what level it has resided within the regolith.

Scientists have also determined with some confidence the age of the Moon and the time of the earliest crystallization of Moon rock formations. Isotopic dating techniques indicate that the earliest crystallization occurred more than 4 billion years ago and that most volcanic activity appears to have spanned some 600 million years.

Methods of Study

Before the first direct contact with the surface of the Moon occurred, much was known about the composition and origin of that surface from sophisticated technology capable of analyzing materials remotely. Those techniques included indirect sensing processes involving the visible, infrared, and microwave bands of the electromagnetic spectrum. The understanding derived from this technology was limited until actual samples of surface materials were brought to Earth by the Moon explorers. Scientists could then compare the data from the remote-sensing experiments with direct observations to determine what it was that they were seeing at the time. Radar, which was used in determining the physical nature of the Moon's surface, was successful in establishing that it was loose, a mixture of sand and broken rock. Infrared studies confirmed that the surface was porous and exhibited low thermal conductivity. Photometric and polarimetric studies further refined the knowledge of the character of the surface by establishing that it contains high concentrations of iron-rich basalts (confirmed by the Apollo missions).

Once the first lunar samples arrived back on Earth, scanning electronic microscopes were used to photograph them, an important step in categorizing individual sample fragments. Samples were also sieved to determine the relative abundance of materials in a particular location. In this way, researchers were able to determine where some materials were more abundant than others on the surface when collected using the same procedure. Some samples were cross-sectioned for better observation of stratification, if it existed. Many were tested using standard chemical analysis techniques to determine the chemical nature of the materials. X-ray fluorescence analyses were also conducted to determine the relative concentrations of specific elements or components. Isotopic and carbon dating techniques were used to determine the relative ages of specific samples and to attempt a chronology of the events that have affected the evolution of the materials under study.

Context

There is enormous scientific interest in the Moon. Many scientists believe that it may hold the key to a better understanding of the physical nature of the solar system—indeed, of the universe itself. Perhaps more important, it may hold the key to a better understanding of the origin of planet Earth and of its relationship to other objects in the universe. Lunar rocks have given scientists the first window into that unknown sphere, helping them to understand the nature of planetary formation. The Luna and Apollo programs demonstrated for the first time that many accepted ideas about the Moon and its composition were accurate, though many others were not. They also demonstrated that the Earth and its satellite have much in common. Except for the lack of a lunar atmosphere of composition similar to that of Earth, they would probably have evolved in much the same way.

Another significant outcome of the study of lunar rock samples has been the unequivocal knowledge that certain elements, metals, and other valuable materials exist in abundance on the Moon and may be important to future generations on Earth. Perhaps more important, however, is the knowledge gained of how specific elements interact under conditions that rarely exist on Earth: namely, with little oxygen and no water. The Moon has now become a giant laboratory that is helping scientists to understand the nature of chemical and geological evolution and order.

In the aftermath of the 2003 Columbia accident, President George W. Bush committed NASA to a Space Vision for Exploration that envisioned the completion of the International Space Station by the second decade of the twenty-first century. At that time the space shuttle fleet would be retired so that humans could resume exploration beyond low-Earth orbit. This plane, which NASA officially named Project Constellation, would have a new crew exploration vehicle (Orion) and lunar lander (Altair) developed to return astronauts to the Moon by 2020, this time to stay and build a permanent base of operations. Constellation efforts were designed to be evolutionary and eventually lead to an expedition to Mars. However, a Moon base would also serve as an outpost from which lunar geology investigations could be organized at a scope well beyond that possible during the Apollo program.

Michael S. Ameigh

Further Reading

Beattie, Donald A. *Taking Science to the Moon: Lunar Experiments and the Apollo Program*. Baltimore: Johns Hopkins University Press, 2003. Explains the science gleaned from the Apollo lunar landings, including the Apollo Lunar Surface Science Experiment Packages (ALSEPs) and their results.

Chaikin, Andrew. *A Man on the Moon: The Voyages of the Apollo Astronauts*. New York: Penguin, 2007. A reissue of one of the most engaging accounts of the Apollo program and the exploration of the Moon. Critically acclaimed, for all ages.

Encrenaz, Thérèse, et al. *The Solar System*. New York: Springer, 2004. A thorough exploration of the solar system from early telescopic observations through space missions that have investigated all the planets. Takes an astrophysical approach, placing our solar system in a wider context as just one of many similar systems throughout the universe.

Hartmann, William K. *Moons and Planets*. 5th ed. Belmont, Calif.: Thomson Brooks/Cole, 2005. An updated version of a classic text that covers all aspects of planetary science. Particularly strong in its presentation of Earth-Moon science. Takes a comparative planetology approach rather than including separate chapters for each planet in the solar system.

National Aeronautics and Space Administration. *Preliminary Science Report: Apollo 11*. NASA SP-214. Washington, D.C.: Government Printing Office, 1969. The first humans on the Moon arrived on this mission and conducted preliminary experiments. Well illustrated and full of graphs.

_____. *Preliminary Science Report: Apollo 12*. NASA SP-235. Washington, D.C.: Government Printing Office, 1970. The results of several experiments on this mission advanced our knowledge of lunar geology significantly. Well illustrated and full of graphs.

_____. *Preliminary Science Report: Apollo 14*. NASA SP-272. Washington, D.C.: Government Printing Office, 1971. This Apollo mission returned significant rock samples from the Fra Mauro region. Well illustrated and full of graphs.

_____. *Preliminary Science Report: Apollo 15*. NASA SP-289. Washington, D.C.: Government Printing Office, 1972. This mission examined the Hadley mountain area and was the first to use the Lunar Rover vehicle. The mission's heat-flow experiments suggested a previously molten interior. Well illustrated and full of graphs.

_____. *Preliminary Science Report: Apollo 16*. NASA SP-315. Washington, D.C.: Government Printing Office, 1972. This mission returned more soil samples and again used the Lunar Rover. Well illustrated and full of graphs.

_____. *Preliminary Science Report: Apollo 17*. NASA SP-330. Washington, D.C.: Government Printing Office, 1973. The final Apollo mission to the Moon landed in the Taurus-Littrow Valley and returned the largest collection of lunar rocks to Earth. Well illustrated and full of graphs.

Schmitt, Harrison J. *Return to the Moon: Exploration, Enterprise, and Energy in the Human Settlement of Space*. New York: Copernicus Books, 2006. A scientific and economic plan for prolonged exploration and exploitation of lunar resources, written by the only geologist to land on the Moon, perform field geology, and collect documented samples for return to Earth. Discusses helium-3 mining, which promises to supply future energy production systems on Earth and thereby finance and systematically expand lunar settlements and research posts.

LUNAR SURFACE EXPERIMENTS

Category: Natural Planetary Satellites

The Moon has been studied from Earth and from lunar orbit, but the most detailed studies must be done on the lunar surface. The United States and the Soviet Union successfully landed eighteen spacecraft on the lunar surface, with many of these missions carrying multiple science payloads.

Overview

The Soviet Union's Luna 1 spacecraft was the first to visit the vicinity of the Moon, flying past it on January 4, 1959. Later missions by both the United States and the Soviet Union crashed into the Moon, collecting photographic and other data on the way. On February 3, 1966, the Soviets' Luna 9 became the first spacecraft to soft-land on the lunar surface. Luna 9, however, had only cameras and a radiation detector as scientific instruments. Luna 13 landed later that year, carrying instruments to measure the density and strength of the lunar regolith (soil) and to study cosmic-ray reflections from the lunar surface. Also in 1966, on June 2, the United States' Surveyor 1 spacecraft soft-landed on the lunar surface. It, too, carried no more than cameras and landing radar with which to study the Moon, but still returned useful data on the nature of the lunar surface. Four other Surveyor probes landed on the Moon over the next two years. All carried cameras, and two carried robotic arms to scrape the surface in order to determine its consistency and to move the regolith into a better position to photograph it. The final three Surveyors also carried alpha-scattering surface analyzers used to

measure the abundances of the many elements making up the lunar regolith.

The Soviet Union successfully landed seven Luna spacecraft on the Moon, with the last one being Luna 24, which touched down August 18, 1976. Three Luna missions (16, 20, and 24) returned small samples of lunar regolith to Earth. Luna 17 and Luna 21 each carried remote-controlled roving vehicles called Lunokhod. From November, 1970, until September, 1971, Lunokhod 1 traveled 10.5 kilometers and returned nearly twenty thousand images. It carried an X-ray spectrometer to study the composition of the lunar soil. Lunokhod 2 traveled 37 kilometers across the lunar surface from January 16 to June 4, 1973, returning about eighty thousand images and performing mechanical tests of the lunar regolith.

The uncrewed lunar missions have proved very important to our understanding of the Moon, but some of the most important science missions were deployed by astronauts during six Apollo landings from 1969 to 1972. Many of experiments were conducted by several of the missions. In addition to conducting experiments on the lunar surface, the Apollo missions collected nearly 382 kilograms of lunar material to bring back to Earth. These samples included rocks as well as samples of the lunar regolith. The landing sites for the Apollo missions were chosen to provide for samples from a variety of different geological features in order to maximize the impact of only a few sample sites. Scientific instruments carried by Apollo 11 in July, 1969, were solar-powered. However, the Moon's slow rotation means that the lunar night lasts for two weeks. The remaining Apollo missions carried radiothermal generators to supplement power for the instrument packages. Instruments for Apollo 12 through Apollo 17 continued operations until the science stations were shut down on September 30, 1977.

All of the Apollo missions carried seismographic equipment with which to study the Moon below its surface. Most of the seismographs were passive systems, monitoring moonquakes. However, Apollo 14 and Apollo 16 both carried active seismographs. These used mortars to fire explosive shells some distance away from the landing site in order to produce seismic waves that could be used to study the lunar geology. A similar seismic system was also deployed by Apollo 17 in December, 1972, except that it used explosive charges carefully placed by the astronauts rather than mortars. In all three cases, the explosives were not detonated until after the astronauts had left the lunar surface.

The Moon and Other Small Bodies

Apollo missions also carried lunar dust detectors designed to study the dust disturbed by liftoff of the lunar lander's ascent stage. Later versions of the detector also included provisions to study the long-term degradation of solar panels exposed to the harsh radiation environment of the Moon. Astronauts also set up foils to capture solar wind particles to return to Earth for study. However, in addition to returning particles for study, ion detectors monitored solar wind and cosmic radiation on the lunar surface, and Apollo 12 deployed a spectrometer to measure the composition of the solar wind. These detectors continued to operate for several years after the end of the Apollo program and returned useful data on variations in solar activity. Ion detectors also were used to monitor gas molecules near the lunar surface. These gases constitute what planetary scientists refer to as the lunar atmosphere. Apollo 17 carried a mass spectrometer to measure the composition of the lunar atmosphere.

Supplementing instruments designed to study the Moon, Apollo 16 carried a far ultraviolet telescope and spectrograph. This was the only astronomical instrument placed on the Moon. Among other targets, it was used to study the Earth and the Large Magellanic Cloud.

To study the geophysics of the Moon, the later Apollo missions also carried magnetometers to study the Moon's residual magnetic field, and Apollo 17 carried an experiment to measure the Moon's surface gravity and to monitor it for any variations over time. Heat-flow experiments to measure the amount of heat flowing from the lunar interior were set up on each of the last three Apollo missions. However, astronaut John Young tripped over the cable for the experiment, breaking it. The cable could not be repaired by the astronauts, so the instrument returned no data.

In addition to experiments carried to the Moon to be performed there, several missions carried mirrored corner reflector arrays that were positioned to point back toward Earth. These corner reflectors, completely passive systems, were designed to reflect light striking them back in the direction from which it came. Powerful lasers fired from Earth at the sites of the reflectors are reflected back to Earth, where they can be detected. Careful measurements of the time that it takes for the light to get to the Moon and back are used to determine the lunar distance and variations in the lunar orbit. The astronauts of Apollo 11 deployed the first corner reflector array. Apollo 14 carried another array, and Apollo 15 deployed a much larger array. In addition to the Apollo arrays, both Lunokhod rovers carried such arrays. With the exception of Lunokhod 1's array, which has not reflected lasers since 1971, these corner reflector arrays are still used today in lunar ranging experiments.

Knowledge Gained

Though the Moon is our nearest neighbor in space, it is still a long way away from astronomers on Earth, and detailed studies of its surface were not possible until the advent of spacecraft capable of traveling to the Moon. One of the first, and very important, steps in studying the Moon was photographic studies of the surface characteristics. Many of the early surface experiments, particularly

Surveyor 1 was the first spacecraft to soft-land on the lunar surface; later Surveyor missions ascertained fundamental surface conditions and paved the way for the Apollo astronauts. (NASA)

those of the Surveyor program, were designed to study the physical properties of the lunar surface to determine if the regolith would be able to support a heavy spacecraft such as the planned crewed missions that were to follow. However, the experimental science stations set up by the Apollo astronauts returned important data about the nature of the Moon itself.

Prior to the surface investigations of the Moon, it was thought that the surface material was largely volcanic in nature. However, the lunar regolith has been found to be composed primarily of impact ejecta from meteorite impacts. Some speculation had been that micrometeorite impacts might have ground the surface of the Moon into a vast ocean of dust, unable to support the weight of a spacecraft landing on it. However that idea was soundly dismissed by the early lunar landings. A layer of fine dust does exist on the lunar surface, but it is more compact than had been thought.

Other surface studies confirmed some ideas put forth after orbital observations that the lunar seas were ancient lava fields that resulted from massive impacts on the lunar surface. Seismological data suggest that the basalts that flooded the impact basins to form the seas did not come all at once, but over a period of time early in the Moon's history. However, a surprising finding was that the impacts that caused these basins occurred primarily near the end of a period of intense bombardment on the Moon rather than randomly distributed in time, as had been suspected.

The experiments performed on the Moon by both crewed and uncrewed missions, along with the studies of lunar samples returned to Earth by the Apollo missions and three Luna missions, have shown the Moon to be an alien world but with some familiarities. The Moon has a much lower density than Earth, and it has a very tiny core. Most of the Moon is composed of material similar to Earth's crust and mantle. Though a few minerals were found on the Moon that did not have counterparts on Earth, most lunar samples were composed of minerals found on Earth. However, the lunar samples were much richer in refractory minerals than Earth rocks, and they contained very few volatile minerals.

Astronomers had assumed that liquid water may have once flowed on the lunar surface, but surface experiments showed a near total lack of water and no indication that water had been present on the Moon. Furthermore, surface investigations suggest that many of the rocks of the lunar highlands formed from a global magma ocean present soon after the Moon's formation.

These findings have revolutionized our understanding of how the Moon and Earth are related to each other. It is now believed, based on the Apollo and Luna findings, that the Moon formed when a giant planetesimal, perhaps the size of the planet Mars, collided with Earth very early in the history of the solar system, perhaps even before the Earth had cooled. The debris from the collision coalesced to form the Moon. Thus, understanding the Moon helps us to understand Earth.

Context

Much of the early research on the Moon was done in the heat of the space race between the United States and the Soviet Union during the 1960's. However, after initial successes, the political will to continue the study of the Moon faded. The last spacecraft to land on the Moon was Luna 24 in 1976. In the early twenty-first century, however, interest in the Moon revived, and several nations developed plans to land spacecraft on the Moon and once again begin studies on the lunar surface.

Experiments to be done on the Moon in the future will build on the work done in the 1960's and 1970's. Scientific theories about the Moon based on the findings of these earlier experiments have led to several theories of the Moon and its evolution, and upcoming missions will carry experiments to test those theories.

Raymond D. Benge, Jr.

Further Reading

Beattie, Donald A. *Taking Science to the Moon: Lunar Experiments and the Apollo Program.* Baltimore: Johns Hopkins University Press, 2001. A well-documented look at the decisions leading to the selection of the lunar science experiments rather than the experiments themselves.

Bond, Peter. *Distant Worlds: Milestones in Planetary Exploration.* New York: Copernicus Books, 2007. This book for the lay reader covers the history of planetary exploration. A chapter is devoted to lunar exploration.

Freedman, Roger A., and William J. Kaufmann III. *Universe.* 8th ed. New York: W. H. Freeman, 2008. An excellent college-level introductory astronomy textbook with a chapter devoted to the Moon. The emphasis is on the understanding of the Moon, though, rather than the experiments leading to that understanding.

Hamblin, W. Kenneth, and Eric H. Christiansen. *Exploring the Planets.* New York: Macmillan, 1990. An excellent overview of planetary geology, with an entire chapter devoted to the Moon. It is well illustrated, though all illustrations are in black and white.

Heiken, Grant, and Eric Jones. *On the Moon: The Apollo Journals*. New York: Springer, 2007. An excellent and well-documented resource for studying the Apollo missions, covering each mission, the process behind selecting the landing sites, and the experiments performed.

Mackenzie, Dana. *The Big Splat: Or, How Our Moon Came to Be*. Hoboken, N.J.: John Wiley & Sons, 2003. An excellent explanation of the current, post-Apollo model of how the Moon formed. One chapter covers science done on the Moon. The book has a glossary and a very good bibliography.

Orloff, Richard W., and David M. Harland. *Apollo: The Definitive Sourcebook*. New York: Springer, 2006. A thorough account of the Apollo missions and results, with a glossary and an extensive bibliography.

METEORITE AND COMET IMPACTS

Category: Impact Events

The effects of meteorite and comet impacts on Earth range from the insignificant to the greatest natural disaster humankind may ever face—the extinction of most of the life on Earth.

Overview

The Moon viewed through even a small telescope is a spectacular sight. It is covered with craters. Samples brought back from the Moon prove that they are impact craters, not volcanic craters. Because Earth and the Moon are in the same part of the solar system, it follows that Earth has been subjected to the same bombardment from space that produced craters on the Moon. Having been largely erased by erosion, Earth's own cratering record is not so obvious. Earth's atmosphere protects it from the rain of smaller meteoroids, a protection the Moon lacks, but the fact remains that Earth has been hit countless times in the past, and no doubt it will be hit countless times in the future.

Objects that are out in space that might hit Earth include dust, meteoroids, asteroids, and comets. In modern terminology, a meteoroid is a natural, solid object in interplanetary space. A meteor is the flash of light produced by frictional heating when a meteoroid enters a planetary atmosphere. Particularly bright meteors are called fireballs or bolides (especially if they explode). Meteorites are meteoroids that survive their passage through the atmosphere and reach the ground.

Photographs of three meteorites during their meteor phase—from Pribram, Czechoslovakia, 1959; Lost City, Oklahoma, 1970; and Innisfree, Alberta, 1977—have allowed pre-impact orbits to be calculated. The orbits of all three were traced back to the asteroid belt. Beginning in 1969, various workers were able to match the spectra of meteorites with those of asteroids, and it is now widely accepted that most meteorites are chips from asteroids. A few have been identified as having come from the Moon or from Mars.

Rocky or metallic objects larger than about 328 feet (100 meters) across are called asteroids. They are so named because they look like stars—like points of light—in a telescope, but they have more in common with planets than with stars. It is believed that when the Sun first formed it was surrounded by a platter-shaped cloud of gases and dust grains. These grains accreted to form ever-larger objects, and the largest ones became the planets. Asteroids and comets are leftover objects that were never incorporated into planets. Asteroids larger than about 18.6 miles (30 kilometers) in diameter contained enough radioactive elements to melt their insides, allowing nickel and iron to sink to the center and stony material to float to the top. Over the eons, collisions among the asteroids have produced the collection present today. Nickel-iron asteroids are the remnant cores of asteroids whose outer, stony material has been chipped away. To penetrate deeply enough into Earth's atmosphere to cause severe damage, objects must be more than about 131, 164, and 328 feet (40, 50, and 100 meters) in diameter for metallic, stony, or icy bodies, respectively.

The main asteroid belt lies between the orbits of Mars and Jupiter. Further out in the solar system, beyond the orbit of Neptune, ice was the most abundant solid building material. (Here, ice means mostly frozen water, but it also includes frozen carbon dioxide, methane, and ammonia.) The solid part of a comet, the nucleus, forms from these ices mixed with silicate and hydrocarbon dust grains. An inactive comet looks much like an asteroid, but as a comet nears the Sun, vapor streams from the nucleus as the ices evaporate. Inactive comets are difficult to detect, but a large, active comet is a spectacular sight. The nucleus is surrounded by a vapor cloud 621,400 miles (1 million kilometers) across and has a gas tail up to 62,140,000 miles (100 million kilometers) long.

Asteroids or comets that may hit Earth are of obvious interest. Richard P. Binzel, a professor at the Massachusetts Institute of Technology, developed a scale to help scientists communicate with the media and the public about the perceived risks associated with these objects. This scale is named the Torino Impact Hazard Scale and was adopted by the International Astronomical Union (IAU) in 1999. A Torino scale 0 object is either too small to cause damage or will not hit Earth. Torino scale 1 objects will probably not hit Earth, but they merit careful watching. Torino scale 2, 3, and 4 objects merit concern, and scale 5, 6, and 7 objects are progressively threatening. Torino scale 8, 9, and 10 objects will hit Earth and are expected to cause local, regional, or global damage, respectively.

Geography

Any place on Earth may be hit by a meteorite; no location is particularly safe, but seacoasts are the most vulnerable. The 1908 Tunguska impact was a Torino scale 8 event with localized destruction. Had the Tunguska meteorite been just large enough to reach the ground intact, the destruction still would have been largely local. However, if such an object struck the ocean it would generate tsunamis that would cause widespread coastal destruction.

The impact of a Tunguska-scale object on the glaciers of Greenland or Antarctica might melt 35,315 cubic feet (1 cubic kilometer) of ice, but that would produce only an imperceptible rise in the ocean level. However, the impact on Antarctica of a 6-mile-diameter asteroid, such as is thought to have killed the dinosaurs, could melt enough ice to raise sea level more than 230 feet (70 meters). Another environmentally sensitive site for a giant impact is a thick limestone deposit such as exists on the Yucatán Peninsula. It seems likely that the copious amounts of carbon dioxide released from the Yucatán limestone contributed to a warmer climate for thousands of years after the impact.

Prevention and Preparations

The first step in meteorite prevention and preparation is to make a survey of objects that come close to Earth. These are called near-earth objects (NEOs). Under the auspices of the International Astronomical Union, the Spaceguard Foundation was established on March 27, 1996, in Rome. The foundation coordinates international efforts to discover NEOs. As of July 29, 1999, 760 NEOs had been discovered and their orbits calculated. The most dangerous of these are 184 potentially hazardous asteroids (PHAs). PHAs are larger than 492 feet (150 meters) in diameter and will come within 4.7 million miles (7.5 million kilometers) of Earth. More refined orbital information should eventually tell whether or not they will actually hit Earth. As of July, 1999, there were no known PHAs with more than a minute probability of hitting Earth.

If it is discovered that an asteroid is about to hit Earth, can anything be done about it? The answer depends upon three key factors: the amount of warning time, the size of the asteroid, and the state of readiness of the space program. Taking the third factor first, there are normally no spacecraft on standby that are capable of reaching an

(NASA)

asteroid. That means that if the warning time is only a few months, the only thing to be done is to evacuate the probable impact site, or to evacuate coastal areas if an ocean impact is predicted. Such an evacuation will be difficult and disruptive for a Torino scale 8 (local damage) object and will approach the impossible for a Torino scale 9 (regional damage) object. It would be incredibly difficult to evacuate the eastern United States, for example. For a Torino scale 10 object (global catastrophe), preparation efforts will be to provide food, shelter, and energy stores to maximize the number of survivors.

Once an asteroid is discovered and observed for a period of time, its orbit can be predicted accurately for fifty to one hundred years into the future. Deflecting the asteroid into a slightly different orbit becomes an option if there is a ten- to twenty-year warning time. Deflection is probably superior to attempting to destroy the object. Objects small enough to be vaporized with nuclear weapons are small enough to be destroyed by Earth's atmosphere. If an asteroid were not vaporized, but rather only shattered, by a nuclear explosion, the cloud of fragments would continue in the asteroid's orbit and still strike Earth. If there were enough fragments, or if there were large fragments, Earth would still be devastated.

Another solution is to explode a nuclear weapon above the surface of the asteroid. Prior experimentation and manned exploration may be necessary to determine how best to do this. Heat and radiation from the blast will vaporize asteroidal surface material, causing it to push against the asteroid like a rocket engine and thereby change the asteroid's orbit. Only a small change in orbit would be necessary if done far enough in advance. A neutron bomb would be the weapon of choice since neutrons would penetrate deeper beneath the surface and therefore launch more material into space than would the gamma rays and X rays of a conventional thermonuclear weapon.

If humankind were to develop sufficient space-faring capacity, workers might land on the asteroid. Given enough time and an energy source such as a nuclear reactor, the orbit of the asteroid could be changed by launching rocks from a catapult device (mass driver) acting as a rocket engine. If there were sufficient water available, as ice in a comet nucleus, or combined in minerals as in some carbonaceous asteroids, steam rockets mounted on the object might be used to change its orbit. Three properties of comets make them more difficult to deal with: The vast majority can be discovered only months before their closest approach to Earth, and they are fragile and may break apart if one tries to maneuver

them. They also travel faster than asteroids. Typical approach speeds relative to Earth are 9.3 miles (15 kilometers) per second for asteroids, but are 15.5 to 31 miles (25 to 50 kilometers) per second for comets.

Rescue and Relief Efforts and Impact

If the damage from a meteorite were local, the aftermath would resemble that of other large-scale disasters, such as massive flooding, large earthquakes, destructive hurricanes, volcanic eruptions, or massive bombings. Rescue and aid workers would come from outside the area, but if a large city were destroyed, it would probably take days to bring sufficient resources to bear. As with the Tunguska event, most impacts occur in sparsely inhabited areas, but such events are expected to occur between once every fifty years to once every thousand years.

If the destruction were regional, it might take many weeks to bring in sufficient aid. During that time the tragedy would be greatly compounded. Such regional events are expected to occur between once every thousand years and once every hundred thousand years. If the destruction were worldwide, sufficient aid would not exist. People in steel and stone buildings might survive the sky becoming baking hot (because of the fiery reentry of debris) unless the air became too hot to breathe or too oxygen-depleted by conflagrations. Both Switzerland and China have large systems of underground shelters built for nuclear war, and many other nations have some shelters. Those who survive the initial impact, earthquakes, tsunamis, hot sky, secondary fires, possibly toxic vapors and gases, and rising sea level (from melting ice) will need food and energy to keep warm for a few months to a year until the worldwide dust cloud settles from the air and the Sun shines again. Then they will need crops that will grow in the new, warmer climate. They will also need to deal with greatly increased ultraviolet radiation from the Sun, plagues, and the breakdown of civilization. Yet, except in an extreme worst case, some people should survive. Global climatic catastrophe due to asteroid or comet impact is expected to occur once every hundred thousand years or less often.

Charles W. Rogers

Further Reading

Burke, John G. *Cosmic Debris: Meteorites in History.* Berkeley: University of California Press, 1986. An engaging treatment of how science discovered the truth about meteorites.

Chapman, Clark R., and David Morrison. *Cosmic Catastrophes*. New York: Plenum Press, 1989. This book treats the K/T impact, in which a meteorite hit the earth 65 million years ago, and other disasters.

Cox, Donald W., and James H. Chestek. *Doomsday Asteroid*. Amherst, N.Y.: Prometheus Books, 1996. A good treatment of the efforts needed to locate and deflect potentially dangerous asteroids and comets.

Lewis, John S. *Rain of Iron and Ice: The Very Real Threat of Comet and Asteroid Bombardment*. Reading, Mass.: Addison-Wesley Publishing Co.: 1996. A good account of various impacts, including interesting, but less well-known, ones.

Sagan, Carl, and Ann Druyan. *Comet*. New York: Random House, 1985. An excellent book by this very successful husband-wife writing team. It explains what we know about comets and how we learned this. The book is easily read and profusely illustrated.

Steel, Duncan. *Rogue Asteroids and Doomsday Comets: The Search for the Million Megaton Menace That Threatens Life on Earth*. New York: John Wiley & Sons, 1995. A good book for the general reader on mass extinctions and the K/T impact, the Tunguska object, and early detection efforts.

Verschuur, Gerrit L. *Impact! The Threat of Comets and Asteroids*. New York: Oxford University Press, 1996. An excellent and authoritative popular work written by an active astronomer.

METEORITES: ACHONDRITES

Category: Small Bodies

Achondrites are a class of stony meteorites containing abundant silicate minerals that have formed as a result of igneous processes on small planetoids or asteroids. They closely resemble basaltic rocks found upon the Earth and Moon.

Overview

Meteorites are solid materials from outside Earth's orbit that have passed through the atmosphere and reached the surface. These objects—made of various combinations of rock and metal—are called meteoroids while still in orbit around the Sun. When a meteoroid encounters the Earth and enters the atmosphere, collisions with air molecules cause it to heat up and begin to vaporize. This produces a glowing trail of superheated air and hot vaporized material from the meteoroid that appears as a flash of light streaking across the sky. This phenomenon is properly called a meteor. (It also is commonly called a "falling star" or "shooting star," but such names are inaccurate since it is not a star at all.) It is estimated that the Earth is bombarded by hundreds of tons of meteoroidal material every day, but most meteoroids are no larger than small pebbles, and they vaporize completely in our atmosphere. If the meteoroid is large enough to survive its fiery plunge and land on the Earth's surface, then it is called a meteorite.

Meteorites are divided into three main groups based on the abundance of metallic and stony minerals they contain: the nickel-iron meteorites (often just called iron meteorites or irons), stony-iron meteorites (or stony irons), and stony meteorites (or stones). Stony meteorites are composed primarily of silicate and oxide minerals with minor amounts of metal. They can be subdivided into two subgroups known as the chondrites and achondrites.

Achondrites are not as common as chondrites; only about 1 in 10 stony meteorites is an achondrite. Achondrites get their name because they lack chondrules, mineral droplets that make up much of the material in chondrites. Achondrites have crystal textures similar to terrestrial igneous rocks, thus indicating that they formed when some larger parent body (perhaps the size of a small planet or large asteroid), melted, differentiated, and then cooled and solidified. Some achondrites have large mineral crystals that resulted from slow cooling as intrusive rocks inside the parent body. Others with smaller crystals were formed by more rapid cooling on or close to the surface of the parent body. Some resemble terrestrial lava flows that are riddled with bubble holes called vesicles, caused by gases that escaped. Some achondrites show evidence of collisions in space, resulting in a rock called an impact breccia that shows the effects of shock metamorphism. The pressure of impact causes the rocks to break apart and the minerals to shatter or deform, while the heat generated causes mineral and rock fragments either to melt slightly or to fuse together, depending upon its intensity.

Most achondrites are rich in one or more silicate minerals such as olivine (a magnesium-iron silicate), pyroxene (an iron-magnesium-calcium silicate), and plagioclase feldspar (a calcium-sodium-aluminum silicate) in varying proportions. Other minerals, such as spinel and chromite (iron-magnesium-aluminum-chromium oxides), are also found, as are small amounts of metal (less than 10 percent) in the form of iron and nickel alloys. In general, achondrites resemble a rock called basalt, which is a very common dark-colored igneous rock found on the Earth and Moon. With a few exceptions, achondrites are older and contain rare isotopes that make them very different from the rocks found on either the Earth or the Moon.

Achondrites can be subdivided into several types based on their texture and chemical composition. First and most abundant are the eucrites. They are similar in appearance to fine-grained terrestrial basalts, and they contain roughly equal amounts of calcium-rich silicate minerals such as plagioclase feldspar and pyroxene. In a hand specimen, a few eucrites exhibit a cumulate texture, which forms by the accumulation of coarse-grained crystals within a magma chamber. Still others possess a vesicular texture (containing many bubble holes or vesicles formed by escaping gases) and closely resemble terrestrial basaltic lava flows. Many eucrites also contain mixed fragments from other meteorite types and show the effects of shock metamorphism. The small variation in abundance of major chemical elements within all eucrites suggests their origin on the same parent body. They probably formed as extrusive and shallow intrusive igneous rocks later blasted out of the parent body by impacts. The age of eucrites has been dated using radioactive rubidium-strontium isotope techniques at 4.5-4.6 billion years, indicating crystallization very early in the history of the solar system.

Diogenites are achondrites that have a chemical composition similar to a terrestrial igneous rock called pyroxenite, which has an abundance of the mineral pyroxene. Texturally, diogenites have coarse-grained crystals that indicate slow, deep cooling below the surface. Based on laboratory melting experiments using actual achondrite samples, these crystals probably formed by cooling and crystallization from the same magma that also produced eucrites or by the more extensive melting of some eucrite source. Chemically, diogenites consist of metamorphosed accumulations of an iron-and-magnesium-rich but calcium-poor pyroxene known as a bronzite, along with minor amounts (less than 10 percent) of plagioclase feldspar crystals and some metallic iron. The bronzite crystals have become chemically homogeneous as a result of metamorphic heating. Like some eucrites, diogenites have been found shattered or mixed with pieces of other meteorites, resulting in a solid rock of angular broken

fragments. Both diogenites and eucrites probably formed on the same parent body.

Howardites are a variety of achondrite that represents mixtures of many different meteorite types. They consist of crushed pieces from eucrites and diogenites, and they also contain about 2-3 percent by weight of pieces from chondritic stony meteorites. Texturally, howardites closely resemble the lunar soil. Under high magnification, a howardite's exterior surface is covered with small micrometeorite craters that contain impact-generated glasses, evidence of their formation by impacts on the surface of the parent body.

A likely parent body for these three types of achondrites seems to be the asteroid 4 Vesta. About 500 kilometers in diameter, it is the only large asteroid with a surface reflection spectrum like that of eucrites and diogenites. Several small Earth-approaching asteroids have similar reflection spectra, and they and the eucrites, diogenites, and howardites found on Earth probably were blasted off of Vesta by one or more large impacts.

Ureilites, another variety of achondrite, were named for the town of Novo Urei in Russia, where the first specimen was found in 1886. They consist of fairly large and abundant crystals of magnesium-rich olivine, some clinopyroxene, and a rare type of plagioclase feldspar set within smaller crystals composed of graphite, iron-rich metals, halite, sylvite, and troilite. In some specimens, the olivine crystals show a preferred orientation from crystal settling while molten; thus, ureilites exhibit variable textures. Most specimens have undergone intense high-pressure shock metamorphism that resulted in the formation of small diamonds from the graphite. Ureilites are the only achondrites that contain these tiny graphite and diamond crystals; the source of the carbon is unknown.

Aubrites, also known as the enstatite achondrites, are composed predominantly of a magnesium-rich pyroxene called protoenstatite and a rare type of plagioclase feldspar. Texturally, aubrites have large crystals, indicating slow cooling, but their origin and place of formation remain unexplained.

SNCs (pronounced "snicks") are a small, unusual, and highly controversial group of related achondrites that include the shergottites, nakhlites, and chassignites. Shergottites are named for the town of Shergotty in the state of Bihar in India, where the first of these strange meteorites fell in 1865. Since that time, a few others like it have been found. As a group, the shergottites are similar to a terrestrial slow-cooled, coarse-grained igneous rock called diabase, which is rich in pyroxene and plagioclase feldspar. One of the feldspars found within the shergottites is maskelynite, a type of feldspar whose orderly atomic lattice structure has become disorganized from shock impact. Other minerals to be found are pyroxenes (calcium-rich augite and calcium-poor pigeonite), calcium- and sodium-rich plagioclase feldspars, oxidized iron in the form of magnetite, some olivine, and a rare water-bearing amphibole named kaersutite. Texturally, the shergottites are cumulates with elongated pyroxene crystals that have a preferred orientation, which is probably the result of flowage of newly formed crystals within the magma while still in a hot liquid state. Their geologic history records crystallization in a relatively Earth-like oxygen-rich environment and a period of intense shock metamorphism and high-intensity heating probably caused by impact, as indicated by numerous quickly cooled glass fragments. The radiometric age determinations on some of the shergottites' minerals reveal a comparatively young age of 1.3 billion years. Trapped gas bubbles within some of the shergottite samples contain nitrogen and noble gases such as xenon, krypton, and argon, very similar in composition to the atmosphere of Mars.

The nakhlites are similar to terrestrial slow-cooled, coarse-grained igneous rocks known as gabbro. Mineralogically, these meteorites contain abundant augite (calcium-rich pyroxene) and smaller amounts of olivine, plagioclase feldspar, a few strange sulfide minerals, and metallic iron. Compared to the shergottites, all these minerals lack shock features, show no evidence of thermal metamorphism, but have a similar radiometric age of 1.4 billion years. Although their overall chemistry is different from that of the shergottites, nakhlites are believed to have been derived from either the same or a similar parent body.

Chassignites are named for Chassigny, France, where the first specimen was found in 1815. Several others have since been found in localities around the world. The few existing samples show that they are composed of abundant crystals of olivine with minor amounts of pyroxene, plagioclase feldspars, and kaersutite that show alteration by shock metamorphism. In a hand specimen, chassignites closely resemble terrestrial olivine-rich rocks called dunite.

Collectively, the SNCs contain cumulate crystals and a large percentage of volatile gases, which indicates formation upon a planet-sized body with a stronger gravitational field than that of the Moon. As a group, they have an average age of about 1.3-1.4 billion years, as determined by radioactive dating techniques. The chemical abundance and isotopic composition of gases trapped within small bubbles in these meteorites is nearly identical to the Martian atmosphere. The SNCs may have formed on Mars, from which they were blasted off into space by large impacts that hit Mars at the correct angle and speed to be eject material at speeds greater than 5 kilometers per second (Martian escape speed); these Martian rocks would eventually have encountered Earth and landed here as meteorites. Specimens from the surface of Mars will first need to be returned to Earth before it is possible to confirm that SNCs really originated there.

In 1974, a single strange meteorite named Brachina was found in Australia. It is similar to the chassignites in mineralogy. Brachina has a fine-grained texture, contains 80 percent olivine and 10 percent plagioclase feldspar, lacks hydrous minerals, is unshocked, and is 4.5 billion years old (much older than any of the SNCs).

Methods of Study

Meteorites arrive daily on the Earth in sizes ranging from specks of dust to huge masses of several thousand kilograms. The vast majority fall into the ocean, never to be recovered, or into remote, uninhabited areas, to be discovered much later, if at all. Most achondrite meteorites probably are passed by unnoticed, because they closely resemble ordinary Earth rocks.

The best place to find meteorites of all types is Antarctica. In this remote, ice-covered continent, meteorites stand out starkly as black rocks against a white background of snow and ice. Because of the extreme cold and lack of liquid water, nearly all varieties of meteorites are found perfectly preserved. Meteorites are usually named for the closest town or post office in the vicinity where they are found; however, in the case of Antarctica, the name of the nearest mountain range, valley, or other topographic feature is used.

In the laboratory, the meteorite is weighed and measured, its density determined, and its physical appearance described. Thin sections are sliced from small chips of the meteorite for viewing under a petrographic microscope, where the behavior of light passing through the individual crystals of the specimen assists in identifying the minerals. The bulk chemical composition of the meteorite and a detailed analysis of its individual minerals can be made using an electron microprobe. When bombarded by an electron beam, the atoms within the specimen emit X rays. The atoms of each element emit X rays with characteristic energies, and the intensity at each X-ray energy indicates the abundance of the corresponding element. The overall texture of the meteorite and the distribution and abundance of each mineral present in it are used to place it in the classification system. Its bulk chemistry and elemental distribution help determine the processes that created it.

Extensive studies suggest that most achondrites probably came from water-free planetoids or asteroids. Based on meteorite melting experiments in the laboratory, the partial melting of a parent body with an overall chondritic composition could produce a eucrite. Other experiments using partially melted igneous rocks containing olivine and plagioclase feldspar produced magmas that, under the proper conditions, could form both diogenites and eucrites. Melting could easily have taken place in the low-pressure environment of space, provided that enough heat was generated by the decay of short-lived radioactive isotopes, such as aluminum 26, that once were abundant in these rocks.

Applying processes similar to those scientists believe formed the Earth, a parent body may be modeled that extensively melted and became partially separated into an upper layer of eucrite material atop a lower layer of diogenite material, surrounding a small, metallic iron-nickel core. The size of this parent body probably was no larger than a few hundred kilometers, so pressures on the interior core region would not have been more than 2 to 3 kilobars (2 thousand to 3 thousand times Earth atmospheric pressure at sea level). The mixing of eucrites and diogenites to form howardites probably occurred via meteorite impact, excavation, and lithification on or close to the surface of the parent body.

Context

Meteorites are samples of the building blocks of the planets. They have provided evidence for reactions in the solar nebula prior to the formation of the planets, processes occurring in planetlike bodies during their formation, and collisional impact events between solar-system objects. Continued study of the achondrites, along with other meteorite types, will provide more clues as to how the planets formed about 4.5 billion years ago.

Steven C. Okulewicz

Further Reading

Bevan, Alex, and John De Laeter. *Meteorites: A Journey Through Space and Time*. Washington, D.C.: Smithsonian Institution Press, 2002. A pictorial work that also deals with radioactive dating and geologic composition of meteorites. For general audiences.

Burke, John G. *Cosmic Debris: Meteorites in History*. Berkeley: University of California Press, 1986. Examines the role of meteorites in science history, from their origin as "thunderstones," their folklore and myths, and their curators and collectors to their role in current research. Many footnotes, a few photographs, and a detailed bibliography. Nontechnical and accessible to the general reader.

Dodd, Robert T. *Thunderstones and Shooting Stars: The Meaning of Meteorites*. Cambridge, Mass.: Harvard University Press, 1986. A thorough, clearly written review of information on all types of meteorites. Chapter 9 discusses achondrites and their parent bodies and how they relate to the origin of the planets.

Hutchison, Robert. *The Search for Our Beginning: An Enquiry, Based on Meteorite Research, into the Origin of Our Planet and Life*. New York: Oxford University Press, 1983. A clear and easy-to-read nontechnical book that describes the various types of meteorites and the information they provide about the solar system. Illustrated with many photographs of meteorites. Highly recommended.

McSween, Harry Y., Jr. *Meteorites and Their Parent Planets*. 2d ed. New York: Cambridge University Press, 1999. Gives an overview of the study of extraterrestrial debris. Good reference for scientists, students, and amateur astronomers.

Norton, O. Richard. *The Cambridge Encyclopedia of Meteorites*. New York: Cambridge University Press, 2002. Thoroughly discusses interior, external, and atomic compositions of meteorites. Valuable resource for scientists, students, and meteorite enthusiasts.

_____. *Rocks from Space: Meteorites and Meteorite Hunters*. 2d ed. Missoula, Mont.: Mountain Press, 1998. Introduces the reader to meteorites, asteroids, comets, and impact craters. This revised and updated edition includes dozens of new photographs.

Norton, O. Richard, and Lawrence Chitwood. *Field Guide to Meteors and Meteorites*. London: Springer, 2008. A guide designed to aid readers in observing meteors, as well as locating and analyzing meteorites.

Reynolds, Mike. *Falling Stars: A Guide to Meteors and Meteorites*. Mechanicsburg, Pa.: Stackpole Books, 2001. A guide for nonscientists on observing meteors and meteorites, including how to photograph them and how to record data.

Spangenburg, Ray, and Kit Moser. *Meteors, Meteorites, and Meteoroids*. Secaucus, N.J.: Franklin Watts, 2002. In addition to providing the basics about its topic, this volume also addresses meteoritic impacts with Earth, the process of watching and hunting meteorites, and what can be learned about the universe from them.

METEORITES: CARBONACEOUS CHONDRITES

Category: Small Bodies

The carbonaceous chondrite meteorites are the most primitive remnants of the primeval nebula from which the Sun, Earth, and all other bodies of the solar system originated. The hydrocarbon molecules present in these meteorites indicate the types of carbon-bearing molecules that most likely were present on the primitive Earth and may have been the building blocks of life.

Overview

Carbonaceous chondrites are a class of stony meteorites that are both chemically and physically primitive. They are chemically primitive in that, except for the elements hydrogen, carbon, oxygen, and the noble gases, the proportions of the elements in these meteorites are very similar to those observed in the Sun. They are physically primitive in that the carbonaceous chondrites escaped the thermal alteration (exposure to heat that causes changes in chemical composition and mineralogy) that affected almost all the meteorites in the other classes. Because of the primitive nature of the carbonaceous chondrites, they are thought to be the best samples currently available of the solar nebula out of which the Sun, Earth, and other solar-system objects formed. Thus, the composition of carbonaceous chondrites is generally taken as the starting point for models of the formation and subsequent evolution of the Earth.

The carbonaceous chondrites are relatively scarce, constituting only about 5 percent of all the meteorites recovered soon after their fall to Earth was actually observed. They are composed of millimeter-sized chondrules, individual grains whose mineralogy and texture indicate their crystallization from molten material, set in a matrix of finer-grained material. The carbonaceous chondrites are easily distinguished from all other meteorites by their dull black color, friability, generally low density, and almost total lack of nickel-iron grains, but not all carbonaceous chondrites are alike. Differences in composition, mineralogy, and texture allow the carbonaceous chondrites to be separated into several distinct types.

The most primitive type of carbonaceous chondrite, called the CI (or C1) type, is extremely rare, represented by only five meteorites. Of these, only Orgueil, a fall of about 127 kilograms, is large enough for extensive study. The others, Ivuna (0.7 kilogram), Alais (with little remaining of a 6-kilogram fall), Tonk (7.7 grams), and Revelstoke (only 1 gram), are all very small. The CI carbonaceous chondrites are different from all other chondrites, both carbonaceous and ordinary, in that they lack chondrules and consist almost entirely of low-temperature minerals, particularly clays.

Another type of carbonaceous chondrite, named CII (or C2 or CM), contains numerous organic compounds, including amino acids, that may have served as the basis for the development of life on Earth. At the very least they provide clues as to the types of organic material likely to have been present on the early Earth to serve as building blocks for life.

The CB carbonaceous chondrites exhibit a high oxidation state, with an abundance of volatile substances. The CO type is slightly less oxidized but contains metal and sulfides. The CV type closely resembles the CO type in mineral composition and oxidation state but contains large quantities of chondrules and whitish aggregates with a high calcium-aluminum content. Together, the CV and CO types are sometimes referred to as the CIII (or C3) type.

The carbonaceous chondrite types other than CI consist of a matrix, similar in chemical composition to bulk CI material, mixed with chondrules and aggregates of minerals that formed at high temperatures from a condensing gas and that exhibit a depletion in volatile elements from what is observed in the CI matrix. As the abundance of high-temperature material increases from about 1 percent in the CI type to about 60 percent in other types, the similarity of the bulk composition to that of the Sun decreases.

At one time, it was thought that the high-temperature material might be derived from the matrix by heating, which would eliminate the volatile material. Recent studies, however, show significant differences between the chemical and isotopic compositions of the high-temperature and low-temperature components of the carbonaceous chondrites, which make it impossible to derive one from the other by any simple process.

Much of the recent research on carbonaceous chondrite meteorites has focused on understanding the process by which these objects formed from the solar nebula, the gas and possibly dust that collapsed to form the Sun, Earth, and other solar-system objects. As the most primitive relics of the formation process currently available for laboratory analysis, the carbonaceous chondrites have been used to determine the chemical composition of the solar nebula, to establish the sequence and duration of events in the formation process, and to determine the temperatures characteristic of the process.

Major advances in the study of carbonaceous chondrites began in 1969. In February of that year, the Allende carbonaceous chondrite fell in northern Mexico, and about 2,000 kilograms of material were recovered for analysis. Later that same year, an even more primitive carbonaceous chondrite, the Murchison, fell in Australia. In addition, the return of lunar samples in 1969 spurred the development of research laboratories for the study of extraterrestrial materials. With the end of the Apollo lunar landing program, many of these laboratories shifted their emphasis to meteorite research and began to use highly sophisticated instruments perfected for lunar sample analysis to study meteorites.

Organic matter (chemical compounds of carbon, nitrogen, and oxygen) has been detected by spectroscopic methods in comets, on some asteroid surfaces, and on some planetary satellites. Study of the properties of this extraterrestrial organic matter would provide indications of the organic material likely to have been present on the Earth at the time life developed on this planet. The meteorites, particularly the carbonaceous chondrites, have been subjected to intensive examinations to determine whether they contain samples of this organic matter.

The search for organic matter in the carbonaceous chondrites was hampered for decades by terrestrial organic contamination of these meteorites from the time of their recovery until their analysis. By the time of the fall of the Murchison meteorite in 1969, researchers were aware of the contamination problem and efforts were made to preserve samples properly for organic analysis. In addition, several laboratories had recently developed procedures and instrumentation to search for organic material in returned lunar samples and to distinguish terrestrial contaminations from extraterrestrial organic material. The first analyses of the Murchison meteorite provided evidence for the presence of amino acids, which are the building blocks of proteins. Subsequent analysis

of carbon in the organic matter indicated that the isotopic composition, the ratio of carbon 13 to carbon 12, was inconsistent with terrestrial contamination. It has now been demonstrated that several carbonaceous chondrites contain a varied suite of organic compounds. These same organic compounds have been duplicated in laboratory experiments by purely chemical processes and consequently are not evidence for life in space, but they are taken to indicate the types and variety of organic material likely to have been present on the Earth to serve as building blocks for life.

Methods of Study

Scientists have employed a variety of techniques and instruments to uncover the secrets locked in the carbonaceous chondrite meteorites. The chemical compositions, molecular abundances, mineralogies, isotopic ratios for individual elements, and present radioactivity have all been studied. Because of the small amount of carbonaceous chondrite material available for scientific study, especially of the rare CI type, many of the techniques employed to examine these meteorites have benefited greatly from the sophisticated instrumentation developed in support of the lunar sample analysis program.

The observation that the carbonaceous chondrites are primitive—that is, relatively unaffected by thermal processes—was established by detailed chemical analyses of individual mineral grains. The effect of prolonged heating is to cause the compositions of minerals of the same type to equilibrate, meaning that all grains of the same mineral from a single meteorite would have approximately the same composition if the meteorite were heated above the equilibration temperature. Such an effect is seen in most ordinary chondrites, but not in carbonaceous chondrites.

The compositions of small mineral grains are usually determined using an electron microprobe, an instrument that bombards the sample with an intense beam of electrons and detects the X rays emitted by the sample. When struck by an electron, each element emits X rays of specific energies. Thus, for example, the number of X rays emitted at the energy characteristic of the element iron gives the iron abundance in the sample. When the mineral grains in carbonaceous chondrite meteorites are examined by this technique, grains of olivine, the most easily altered of the major minerals in the matrix, exhibit a relatively wide range of compositions. In the Allende meteorite, for example, magnesium-rich olivine chondrules are found in direct contact with iron-rich matrix olivine. Such contacts eliminate the possibility that significant thermal events have occurred since the time at which the Allende meteorite was formed.

Observation of radioactive effects can also indicate the thermal history of these meteorites. The elements uranium and plutonium decay by nuclear fission, a process by which the nucleus splits into two fragments, each about one-half the mass of the original nucleus. These fragments fly apart with high energy, traveling a few thousandths of a centimeter before coming to rest. The host material can be damaged along the path of each fragment. This damage, called a fission track, is more easily attacked by reactive chemicals than is the surrounding mineral. After chemical etching, the fission tracks can be observed through a microscope. This damage, however, can be healed by heating. Thus, if fission tracks are revealed by chemical attack, the mineral has not been heated above the healing, or annealing, temperature after the fission event. The presence of tracks from uranium and plutonium fission in minerals from the carbonaceous chondrites indicates that they have not been heated above a few hundred degrees Celsius since their formation.

Similar radioactive decay processes can be used as clocks, providing a way to determine the ages of these meteorites. One such clock depends on the radioactive decay of rubidium 87, an isotope of the element rubidium, into strontium 87, one of the four stable isotopes of strontium. One-half of any initial sample of rubidium 87 decays to strontium 87 in 47 billion years. In any given sample, if the abundance of rubidium 87 as well as the amount of strontium 87 produced by radioactive decay could be measured, the elapsed time, or age, required for that amount of decay could be determined. In practice, application of this radioactive clock is complicated by a number of factors, including the migration of the strontium 87 from its decay site because of heating and the fact that not all the strontium 87 in the sample is from rubidium 87 decay. When appropriate corrections are made for these effects, however, the rubidium-strontium clock, as well as similar decay clocks using other pairs of elements, gives a consistent picture that the carbonaceous chondrites formed 4.55 billion years ago. Results for the oldest rocks on the Moon give essentially the same age. The extensive thermal activity in the early history of the Earth has apparently destroyed most or all evidence of the earliest rocks to form on this planet, but the observation that both the meteorites and the oldest

lunar rocks have a common age suggests that the entire solar system, including the Earth, formed at that time.

Radioactive elements also provide clues to the duration of solar-system formation. Decay of aluminum 26, which is reduced to one-half of its starting abundance in only 720,000 years, produces magnesium 26. Magnesium has three stable isotopes, which are usually found in fixed ratios to one another, but the ratio of magnesium 26 to the other two isotopes of magnesium will increase when aluminum 26 decays. In some of the high-temperature aggregates from the Allende meteorite, significant enrichments in magnesium 26 were found by mass spectrometry. Detailed examination of the minerals containing these enrichments showed that the size of the magnesium 26 enrichment increased in proportion to the aluminum concentration in that mineral. This suggested that radioactive aluminum 26 was incorporated into the mineral and subsequently decayed to magnesium 26. For this to be true, however, the high-temperature aggregates would have to have formed within a few million years of the isolation of the solar nebula, or most of the aluminum 26 would already have decayed. Thus, the high-temperature aggregates in Allende and some other carbonaceous chondrites provide evidence that mineral grains condensed very early in the solar-system formation process.

Context

As a group, the carbonaceous chondrites have provided a wealth of information on many different aspects of planetary formation and the development of biological processes that may have occurred in the early solar system. They exhibit a variety of conditions of formation that range from a high-temperature, low-pressure, volatile-poor environment to one that was at a lower temperature and volatile-rich. They appear to have experienced only minor alteration since their formation from the collapsing gaseous nebula that became our solar system, and thus they preserve a record of that early era of solar-system history. The chemical composition of the least altered of these meteorites is almost identical to the composition of the Sun, except for a few gaseous elements. Thus, the carbonaceous chondrite composition is taken to indicate the bulk composition of the Earth, which cannot be measured directly since the Earth's interior is inaccessible.

Radioactive clocks in the carbonaceous chondrites indicate that they formed 4.55 billion years ago. The consistency of this age with the age of the oldest rocks brought back from the Moon by the Apollo lunar landings is taken to indicate that the entire solar system, including the Earth, formed at that time. Isotopic relics of other radioactive elements, now extinct, demonstrate that some minerals in the carbonaceous chondrites formed within as little as a few million years of the isolation of the solar nebula from addition of new galactic radioactive isotopes.

The carbonaceous chondrites also contain organic molecules, including amino acids, which are the building blocks of proteins. Although there is no evidence of biological activity on the parent body of the carbonaceous chondrites, these or similar organic molecules are likely to have been available on Earth to serve as building blocks for the development of life.

George J. Flynn

Further Reading

Dodd, Robert T. *Meteorites: A Petrologic-Chemical Synthesis*. London: Cambridge University Press, 1981. This is a well-illustrated summary of the mineralogical and chemical analyses of all types of meteorites. Chapter 2 focuses on the chondritic meteorites and their relation to one another. Chapter 3 describes the properties of the carbonaceous chondrites, their relationship to the solar nebula, and the possibility that they contain presolar grains. Text is suitable for college-level readers who have a minimal background in Earth science.

_____. *Thunderstones and Shooting Stars: The Meaning of Meteorites*. Cambridge, Mass.: Harvard University Press, 1986. This reference book explains why there is a scientific interest in meteorites and summarizes what is known about them. It includes chapters on the chondritic meteorites and on the parent bodies of these meteorites. While written as a college-level text, it provides detailed explanations of the phenomena and techniques of analysis without requiring the reader to have an Earth science background.

Erickson, Jon. *Asteroids, Comets, and Meteorites: Cosmic Invaders of the Earth*. New York: Facts On File, 2003. Part of The Living Earth paperback series. Discusses the threats that these objects can pose to Earth and the devastating effects that major impacts have had, and still pose, to life on Earth. Also provides a general description of these bodies separate from their threat of impacting the planet. For the general reader.

Hutchison, Robert. *The Search for Our Beginning: An Enquiry, Based on Meteorite Research, into the Origin of Our Planet and Life*. Oxford, England: Oxford University Press, 1983. Summarizes the present state of scientific knowledge about the Earth, the Moon, and the inner planets and describes how the knowledge gained from the study of meteorites has shaped theories of the origin and evolution of the inner solar system. Illustrated; intended for general readers.

Kerridge, John F., and Mildred S. Matthews, eds. *Meteorites and the Early Solar System*. Tucson: University of Arizona Press, 1988. A collection of articles by sixty-nine contributing authors describing the state of meteorite research as it relates to early solar-system processes. Well illustrated; includes a detailed review and comprehensive bibliography of the major topics in meteorite research accessible to general-level readers. Aimed at graduate students.

Mason, Brian. *Meteorites*. New York: John Wiley & Sons, 1962. This book emphasizes the chemical and mineralogical measurements made on the meteorites and describes the relationship between meteorites and other objects in the solar system. It provides a state-by-state listing of all the meteorites collected in the United States up to the date of the book's writing, including the exact location, year of recovery, weight, and type.

Nagy, B. *Carbonaceous Meteorites*. New York: Elsevier, 1975. This book describes the carbonaceous chondrite meteorites, with particular emphasis on the carbon-rich phases, and discusses the long effort to identify organic compounds in meteorites. Intended as a college text, this book is suitable for readers with a high school science background.

Wasson, John T. *Meteorites: Classification and Properties*. New York: Springer, 1974. A college-level introduction to meteorite research and classification by type. The emphasis is on interpretation of the chemical and mineralogical data. Well illustrated; includes a tabulation of all meteorites known by the early 1970's, with the type of each meteorite indicated. Appendix C lists each carbonaceous chondrite known at the time of publication.

_____. *Meteorites: Their Record of Early Solar-System History*. New York: W. H. Freeman, 1985. This college-level text is less technical than Wasson's 1974 book. This well-illustrated volume describes the formation processes for the different meteorite types and attempts to link the different meteorite groups with their appropriate parent bodies. Chapter 7 provides an extensive discussion of the chondritic meteorites, including the carbonaceous chondrites.

METEORITES: CHONDRITES

Category: Small Bodies

Chondrites are a class of stony meteorites that contain chondrules, small mineral droplets that make up a prominent part of their material. Chondrites are the most common type of meteorite.

Overview

Meteorites are divided into three main groups based on the abundance of metallic and stony minerals they contain: the nickel-iron meteorites (often just called iron meteorites or irons), stony-iron meteorites (or stony irons), and stony meteorites (or stones). Stony meteorites (or stones) are the most abundant of the three groups. They are composed mostly of the silicate minerals olivine, pyroxene, and plagioclase feldspar. Metallic nickel-iron grains occur in varying small amounts and are accompanied by an iron-sulfide mineral called troilite, which is very rare on Earth. Stony meteorites have the greatest variety in composition, color, and structure.

One particular structural feature called chondrules divides the stony meteorites into two main subgroups: the chondrites, those with chondrules, and achondrites, those without. Chondrites are much more common than achondrites; about nine of ten stony meteorites are chondrites. The words "chondrule" and "chondrite" are derived from the Greek word *chondros*, meaning grain. Chondrules are small, rounded particles generally made of high-temperature silicate minerals whose texture indicates rapid crystallization from molten material. These chondrules range in size from less than a millimeter to just under a centimeter. They can be either whole or partial and are embedded within a matrix of fine-grained opaque minerals and glass. The most common types of chondrules are the barred olivine, the excentroradial "feathery" pyroxene, and the porphyritic olivine and pyroxene varieties. These mineral textures, along with the interstitial glass, are indicative of rapid cooling from a high-temperature liquid.

Mineralogically, most chondrites typically are composed of about 45 percent olivine, 25 percent pyroxene, 10 percent plagioclase feldspar, 5 percent troilite, and 2 to 15 percent nickel-iron alloy minerals (kamacite and taenite). Chondrites that tend to be poorer in olivine have a correspondingly higher metal content, and the reverse is also true. This relationship gives rise to a classification scheme in which the three main groups of chondrites are defined: the H type (high iron content), the L type (low iron content), and the LL type (very low iron content). Visually, specimens may

be roughly classified into one of these groups by the amount of metal observed on a cut surface. Texture also can play a part in identifying specific types, as the LL type tends to be more brecciated (composed of rock fragments) than the H and L types. These three groups make up the vast majority of chondrites, but other, rarer groups are also recognized. The E type (or enstatite) chondrites are named for the predominance of the mineral enstatite (an iron-free magnesium-rich pyroxene); iron, constituting about 15 to 25 percent of this type's composition, commonly occurs in the metallic state. The principal differences in the mineralogy of these chondrite types can be attributed to different oxidation states and thus reflect specific conditions of formation. The H, L, and LL chondrites are indicative of a relatively high oxidation state. The E chondrites, on the other hand, are indicative of more reducing conditions and offer an interesting insight into variations in meteorite formation processes.

Another rare group, the carbonaceous chondrites (C type), derives its name from the carbon found in their bulk chemistry, which far exceeds the normal trace amounts found in the ordinary chondrites. Most chondrites (the ordinary chondrites) contain little or no carbon, but a few (the C type, or carbonaceous chondrites) contain a suite of carbon compounds. The carbonaceous chondrites exhibit a wide variation in their chemistries and conditions of formation. These differences divide them into several distinct types. The most primitive type of carbonaceous chondrite, called the CI (or C1) type, is different from all other chondrites, both carbonaceous and ordinary, in that carbonaceous chondrites lack chondrules and consist almost entirely of low-temperature minerals, particularly clays. The CII (or C2 or CM) type contains numerous organic compounds, including amino acids, that may have served as the basis for the development of life on Earth. The CB carbonaceous chondrites exhibit a high oxidation state, with an abundance of volatile substances. The CO type is slightly less oxidized but contains metal and sulfides. The CV type closely resembles the CO type in mineral composition and oxidation state but contains large quantities of chondrules and whitish aggregates with a high calcium-aluminum content. Together, the CV and CO types are sometimes referred to as the CIII (or C3) type. The carbonaceous chondrites exhibit a variety of conditions of formation that range from a high-temperature, low-pressure, volatile-poor environment to one that is at a lower temperature and volatile-rich. As a group, the carbonaceous chondrites have provided a wealth of information on many different aspects of planetary formation and the development of biological processes that may have occurred in the early solar system.

Chondrites show varying degrees of similarity to Earth rocks, yet they are distinct from any Earth rock. Their basic mineralogy reveals that most chondrites were originally formed under high-temperature conditions from molten material like igneous rocks but were later broken up and reassembled, like sedimentary rocks. Once incorporated into a new mass, the chondrites were subject to variations in temperatures and pressures, thus becoming metamorphosed. It is evident that they have had a rather complex evolutionary history and can therefore provide interesting evidence for the interpretation of early solar-system history.

The study of chondritic meteorites therefore raises important questions based on variations in their chemical compositions and textural features. How did they condense from the solar nebula, and what type of parent body produced all these variations? To try to provide some answers, a hypothetical parent body between 200 and 300 kilometers across is proposed. This body would have been chondritic in nature and subjected to partial melting, presumably because of the short-lived radioisotope aluminum 26. Later bombardment by smaller bodies would produce localized melting and the brecciation that is common to most chondrites. This model is highly speculative but does offer a reasonable explanation for many of the features of the different types. In addition, some of the chondrites retain some of their pre-parent-body characteristics and offer evidence of the conditions that existed before the accretion of asteroid-sized bodies. Thus the chondritic meteorites play an important part in the understanding of planetary formation.

Methods of Study

Chondritic meteorites are studied by many different analytical methods. The first is the determination of bulk chemical content. This can be achieved through basic wet chemical techniques or by the use of more sophisticated methods such as neutron activation analysis and X-ray fluorescence. Where individual minerals are large enough, X-ray diffraction can be employed for a positive identification of the particular mineral phase. However, the mineral grains in most chondrites usually are too small and require an alternative approach using an electron microprobe. This instrument utilizes a microscope to locate tiny individual mineral grains, and then electron bombardment of the grain causes the atoms of each element in it to emit X rays with characteristic energies, making it possible to determine very precise mineral phases at the microscopic level.

Once the meteorite's mineralogy is known, it becomes possible to employ radioisotope dating techniques to learn the age at which the minerals crystallized. This can be achieved by use of potassium-to-argon (K-Ar), rubidium-to-strontium (Rb-Sr), uranium-to-lead (U-Pb), and thorium-to-lead (Th-Pb) radioactive decay rates. It is from such data that some chondritic meteorites have been established as the oldest known solid materials in the solar system.

To determine what has physically happened to the chondrite over the eons requires optical analysis with a petrographic microscope. This instrument offers a magnified view of the texture of the meteorite and the minerals it contains, which in turn provides a look at features that relate to the original condition of the meteorite and the significant changes that occurred at later dates. In this technique, a thin slice (about 0.03 millimeter thick) of the meteorite is cut and adhered to a glass plate, thus permitting light to pass through. In this way, researchers can easily identify mineral grains, examine the nature and appearance of any chondrules, and look for evidence of metamorphism and the characteristics of thermal and mechanical alteration resulting from shock impact.

Experimental petrology—laboratory experiments designed to reproduce the mineralogies and textures found in chondrules—has provided valuable data for developing theories to explain chondrule formation and the accretion process of the chondrites themselves. Mathematical models and computer simulations also are used to try to explain chondrite origin. All these different techniques provide clues about the conditions and processes of meteorite formation and their relationship to planetary formation.

Context

Meteorites in general provided humans' first contact with extraterrestrial materials. Chondrites are similar to certain types of Earth rocks in both their chemistry and their mineralogy. The chondrules they contain have proven to be the oldest known solid material in the solar system. Detailed examination of these chondrules and their matrix material reveals evidence of the processes leading to the formation of the planets.

The carbonaceous chondrites provide information about the organic chemistry that developed in space and the origin of the compounds that eventually may have led to the beginning of life on Earth. The study of chondritic meteorites has revealed much about the conditions that existed when the solar system was formed. They indicate a set of unique conditions that may have lasted only a short period of time and therefore may be the key to understanding how the terrestrial planets came to exist. They have given us new perspectives about our origins and our home, the Earth.

Paul P. Sipiera

Further Reading

Bevan, Alex, and John De Laeter. *Meteorites: A Journey Through Space and Time*. Washington, D.C.: Smithsonian Institution Press, 2002. A pictorial work that also deals with radioactive dating and geologic composition of meteorites. Good reference work. For general audiences.

_____. *Thunderstones and Shooting Stars: The Meaning of Meteorites*. Cambridge, Mass.: Harvard University Press, 1986. A very good introduction to the science of meteoritics at a very basic level. It is a good review of the chemical types and methods of study used to classify meteorites. In addition, there is some discussion about the importance of meteorites as a planet-shaping process and about the effect they may have on life-forms throughout the ages. It is best suited for a reading level of high school to college.

Hutchison, Robert. *The Search for Our Beginning: An Enquiry, Based on Meteorite Research, into the Origin of Our Planet and Life*. Oxford, England: Oxford University Press, 1983. A well-written introduction to meteorites and their relationship to planetary formation. Well illustrated; somewhat technical but suitable for high school and college-level readers.

Mason, Brian. *Meteorites*. New York: John Wiley & Sons, 1962. Perhaps the best of the early books on meteorites and their importance to science. Although this book is dated in the light of modern technology, it still remains an excellent primer for the study of meteorites. It is well written and is best suited for high school and college readers.

McSween, Harry Y., Jr. *Meteorites and Their Parent Planets*. 2d ed. New York: Cambridge University Press, 1999. Gives an overview of the study of extraterrestrial debris. Good reference for scientists, students, and amateur astronomers.

Norton, O. Richard. *The Cambridge Encyclopedia of Meteorites*. New York: Cambridge University Press, 2002. Thoroughly discusses interior, external, and atomic compositions of meteorites. Valuable resource for scientists, students, and meteorite enthusiasts.

_____. *Rocks from Space: Meteorites and Meteorite Hunters*. 2d ed. Missoula, Mont.: Mountain Press, 1998. Introduces the reader to meteorites, asteroids, comets, and impact craters. This revised and updated edition includes dozens of new photographs.

Norton, O. Richard, and Lawrence Chitwood. *Field Guide to Meteors and Meteorites*. London: Springer, 2008. This guide is designed to aid those who observe meteors, including locating and analyzing meteorites.

Reynolds, Mike. *Falling Stars: A Guide to Meteors and Meteorites*. Mechanicsburg, Pa.: Stackpole Books, 2001. A guide for nonscientists on observing meteors and meteorites, including how to photograph them and record data.

Spangenburg, Ray, and Kit Moser. *Meteors, Meteorites, and Meteoroids*. Secaucus, N.J.: Franklin Watts, 2002. In addition to covering the basics, this book covers impacts with Earth, how to watch and hunt meteorites, and what can be learned about the universe from them.

Wasson, John T. *Meteorites: Their Record of Early Solar-System History*. New York: W. H. Freeman, 1985. A well-written and well-illustrated introduction to the science of meteoritics. The author covers most of the significant topics in a clear and understandable way and offers a wealth of information for both the casual reader and the serious student. Suitable for high school and college levels.

METEORITES: NICKEL-IRONS

Category: Small Bodies

Nickel-iron meteorites (often simply called iron meteorites) are one of the three main groups of meteorites. The nickel-iron group has an approximate composition ratio of more than 80 percent metals to less than 20 percent stony material. The metal in them mostly is iron, with nickel present in much smaller amounts.

Overview

Meteorites are objects of extraterrestrial origin that have intersected the orbit of the Earth, survived passage through the atmosphere, and reached the Earth's surface in various stages of preservation. Mineralogically, meteorites may contain various proportions of nickel-iron alloys, silicates, sulfides, and various other minor minerals. They are broadly classified into three major groups: nickel-iron meteorites (often called iron meteorites or just irons), stony-iron meteorites (or stony irons), and stony meteorites (or stones). This classification is based on the ratio of metallic to stony minerals. The irons generally contain more than about 80 percent metals, the stony irons have about a 50-50 ratio, and the stones generally contain more than 80 percent stony minerals.

Stony meteorites are by far the most common, accounting for approximately 95 percent of the meteorites observed to fall to Earth and the large number of meteorites that have been collected in Antarctica. Iron meteorites once were thought to be much more common than they actually are, since most meteorites that had been found (not just those observed falling) were irons. This was because iron meteorites are more easily noticed on the ground, standing out more distinctly from terrestrial rocks than the other types do. Collecting meteorites in Antarctica, where all types stand out prominently against a white background of snow and ice, has now shown that irons account for only about 3 to 4 percent of recovered meteorites. Stony irons are even rarer, accounting for no more than 1 percent of the total.

Iron meteorites are composed of nickel-iron alloy minerals that occur in the metallic state. There is no native terrestrial equivalent for these minerals, and in fact the only native metallic iron found on Earth is in small amounts on Disko Island, Greenland, and in Josephine, Oregon. The most common form in which terrestrial iron is found is in the oxide state, in the minerals hematite, magnetite, and limonite. In contrast, the conditions under which meteoritic iron formed were oxygen-poor. This absence of oxygen, combined with the percentage of nickel alloyed with iron, indicates an extraterrestrial origin.

Iron meteorites, or siderites as they were once called, are characterized by the presence of two nickel-iron alloy phases consisting of kamacite ($Fe_{93}Ni_7$) and taenite ($Fe_{65}Ni_{35}$), combined with minor amounts of troilite (FeS) and other rare mineral phases. Based on the percentage of nickel to iron present, iron meteorites are divided into three subgroups: hexahedrites, octahedrites, and ataxites. Hexahedrites possess a bulk chemical composition of 4-6 percent nickel, occurring principally in large single crystals of the mineral kamacite. Octahedrites, which are the most common, contain increasing amounts of nickel, appearing in the mineral form of taenite along with kamacite. The third group, the ataxites, has a nickel content in excess of 18 percent, with taenite and

an intergrowth mixture of kamacite and taenite called plessite present.

The two nickel-iron alloy minerals kamacite (up to 7.5 mass percent nickel) and taenite (between 20 and 50 mass percent nickel) are the two most abundant minerals in iron meteorites. More than forty other minerals have also been identified but are present in only minor amounts. Among these minerals, troilite, diamond, and graphite are the most significant. The others have no terrestrial equivalent and have been reported only from meteorite studies.

The mineralogy of iron meteorites is unique also in textural appearance as a result of the relationship between the coexisting kamacite and taenite during the meteorite's cooling process. A mixture of kamacite and taenite produces a geometric pattern of intersecting crystals called Widmanstätten structure, named for its discoverer, Alois Josep Widmanstätten (1754-1849), director of the Imperial Porcelain Works in Vienna. This weave-like or crosshatched pattern is revealed when a cut surface of the iron meteorite is polished and then etched with nitric acid. The pattern results from plates of kamacite occurring in octahedral orientation with the spaces in between filled with taenite. The bandwidth of the pattern depends on the width of the kamacite plates, which varies according to their nickel content. The pattern is thought to

This meteorite, found by the Mars Exploration Rover Opportunity on the Martian surface, was nicknamed Heat Shield Rock and consists mostly of nickel and iron. (NASA)

be the result of slow cooling over millions of years while the iron resided inside a small asteroid-sized body. The Widmanstätten pattern does not occur in any known terrestrial rock; it is an important criterion in positively identifying a piece of iron as an iron meteorite.

The formation of the three subgroups of iron meteorites is directly related to the amount of nickel originally present, falling temperatures, and the resulting rearrangement of iron and nickel atoms; each subgroup was produced as a certain temperature was passed. The process began as the temperature fell below about 1,700 kelvins (1,400° Celsius), allowing taenite to crystallize. When the temperature dropped below about 1,120 kelvins (850° Celsius), diffusion of nickel occurred, and the crystal structure of the taenite readjusted to accommodate the formation of kamacite. That was possible because both minerals have crystal structures with cubic symmetry, but the size difference between nickel and iron atoms gives each mineral a different crystal form. Kamacite has a "body-centered" crystal lattice; each atom is found at the center of a cube and is surrounded by eight neighboring atoms. In contrast, taenite has a "face-centered" crystal lattice, with an atom centered on each face of a cube; each atom is surrounded by twelve neighboring atoms. The packing arrangement of the atoms in taenite is more compact, thus allowing it to fill the spaces between the kamacite plates.

The study of cooling rates as determined for numerous iron meteorites reveals a wide range. This finding implies that they originated at several different depths rather than in a single core, as once thought. If so, the parent body would have been relatively small (probably between 100 and 300 kilometers in diameter) and would have had a mass insufficient to melt its interior totally. Partial melting could have taken place as a result of radioactive heating as isotopes such as aluminum 26 decayed; this could have created pockets of molten nickel-iron randomly scattered throughout the parent body. Later impacts with similar-sized bodies could have freed them to assume independent orbits as relatively pure lumps of metal alloys. The shock deformation lamellae (called Neumann lines) seen in the hexahedrites may be evidence of such events.

Methods of Study

Field recognition of a meteorite is not an easy task, unless one is very familiar with its distinctive characteristics. Usually the most obvious feature of a meteorite will be its unusual heaviness as compared to terrestrial rocks of similar size. This is especially the case for iron meteorites, since they are generally about three times denser than typical Earth rocks. Another easily testable property of an iron meteorite is its strong attraction to a magnet. The surface of a meteorite is fairly smooth and featureless but will often exhibit flowlines, furrows, shallow depressions, and deep cavities. One very characteristic surface feature is shallow depressions known as thumbprints, because they resemble the imprints of thumbs pressed into soft clay. Newly fallen meteorites can also exhibit a fusion crust, which shows the effects of intense atmospheric heating upon its surface. In appearance, this crust resembles black ash, but it will weather to a rusty brown and even disappear with time. The fusion crust on iron meteorites is not particularly distinctive and does weather rapidly.

In most cases, positive confirmation of an iron meteorite must be made in the laboratory. A small corner of the specimen can be cut, polished, and etched with acid to look for Widmanstätten patterns. If these patterns are found, then the specimen is definitely an iron meteorite, but not all iron meteorites show Widmanstätten structure. A relatively simple chemical test for the presence of nickel can be made by dissolving a small amount of the specimen in hydrochloric acid; then tartaric acid, 1 percent solution of dimethylglyoxime in ethanol, and ammonium hydroxide are added. If the solution contains nickel, a scarlet precipitate will result. A quantitative analysis is conducted then to determine the actual mass percentage of nickel. Because nickel content in meteorites falls within a very specific range, this determination will confirm the sample's identity.

Over the years, various criteria have been used to classify iron meteorites. Some of the more obvious have been chemical, structural, and mineralogical; others include cosmic-ray exposure ages and cooling rates. A widely used system, which goes back to the late 1800's, is based on the bandwidth of the Widmanstätten structure (the octahedral array of kamacite) as seen on a cut, polished, and etched surface. The width of these bands of kamacite is dependent on both nickel content and cooling rates. Bulk nickel content generally increases as the bandwidth of kamacite decreases, thus providing a criterion for assigning individual specimens to common groups. Chemical studies for trace elements have extended this classification scheme by including analyses for gallium and germanium. A good correlation has been found between bandwidth size and gallium content, thus permitting a finer separation of iron meteorites into smaller subgroups.

Studies that classify the irons into specific types also provide clues to the meteorite's origin and the nature of

About twenty thousand years ago, near Winslow, Arizona, an iron meteorite struck Earth, creating what is now known as Meteor Crater, 200 meters deep and 1 kilometer in diameter. (D. Roddy, U.S. Geological Survey/Lunar and Planetary Institute)

its parent body. An estimation of the cooling rate for the coexisting kamacite and taenite can provide evidence of conditions at the time of the meteorite's origin. This cooling rate has been determined from crystallization experiments in the laboratory and from direct observation of the mineral phases found in iron meteorites. The estimated cooling rates vary with the bandwidth sizes of the kamacite phase, and this provides a correlation between cooling rates and bulk chemical composition.

The determination of the cosmic-ray exposure age of a meteorite indicates when the object broke out of its parent body. This technique may also lead to the matching of individual meteorite specimens to a common event. In addition, the compositions and abundances of minor and trace minerals, along with the extent of shock damage to their structures, might give a clearer picture of the events that led to and occurred during the parent body's breakup. Studies such as these reveal clues not only about the origin of the meteorite but also about the formation of the Earth.

Context

Nickel-iron meteorites had an effect on early human history and technology. Some of the earliest historical records from ancient Egypt speak of iron falling from the sky, and it was undoubtedly meteoritic iron that was first fashioned into iron tools and weapons. Studies have shown that iron tools manufactured on South Pacific Islands, where no local source of iron could be found, were actually forged from meteoritic iron. Some ancient cultures also worshiped "heavenly" iron and placed it in the burial tombs of their leaders; it was thought to be a gift from the gods and served as a symbol of wealth and power. In Europe, as the Bronze Age ended, iron actually became more valuable than gold. Perhaps in the not-too-distant future, space colonists will be mining iron asteroids to provide for their industrial needs.

Today, nickel-iron meteorites are helping reveal the processes of planetary formation in the early solar system. They provide evidence of what the interiors of the terrestrial planets may be like. The Earth's core probably is composed of a nickel-iron alloy similar to that found in iron meteorites.

The scars of giant impacts, many due to iron meteorites, dot the Earth's surface from Arizona to Australia. Perhaps the most recent testimony to the effects of a giant meteorite impact can be seen at Meteor Crater near Winslow, Arizona. At that site, more than twenty thousand years ago, an iron meteorite weighing more than 100,000 tons collided with the Earth. The resulting crater, which measures more than 1 kilometer across and nearly 200 meters deep, was created by an object about 30 meters across traveling at a speed of 15 kilometers per second. The energy released at impact was on the order of a 2- or 3-megaton nuclear weapon, destroying most of the meteorite in the process, but that it was an iron meteorite has been confirmed because broken fragments and solidified droplets of meteoritic iron have been recovered around the site.

The largest known intact meteorite is an iron meteorite, the Hoba West meteorite, which weighs an estimated 60 tons and is still embedded in the ground where it fell near Grootfontein in Namibia, southwestern Africa. The largest meteorite on display in a museum is another iron meteorite, the Ahnighito meteorite, which weighs 34 tons; it was found by the arctic explorer R. E. Perry near Cape York, Greenland, in 1894, and brought to New York for display in the American Museum of Natural History.

Paul P. Sipiera

Further Reading

Bevan, Alex, and John De Laeter. *Meteorites: A Journey Through Space and Time.* Washington, D.C.: Smithsonian Institution Press, 2002. A pictorial work that also deals with radioactive dating and geologic composition of meteorites. Good reference work. For general audiences.

Buchwald, Vagn F. *Handbook of Iron Meteorites: Their History, Distribution, Composition, and Structure.* Berkeley: University of California Press, 1975. This three-volume work offers both an excellent introduction to the science of meteoritics and a general reference to specific iron meteorites. For the average reader, it provides all the basics on meteorites' origin and chemical nature. For the scientist, it provides the best possible reference source for individual specimens. Most suited for college and graduate levels.

Dodd, Robert T. *Thunderstones and Shooting Stars: The Meaning of Meteorites.* Cambridge, Mass.: Harvard University Press, 1986. A very good introduction to the science of meteoritics at a very basic level. Reviews the chemical types and methods of study used to classify meteorites and makes reference to the importance of meteorites as a planet-shaping process and to the effect they may have had on life-forms throughout the ages. Suitable for high school and college students.

Hutchison, Robert. *The Search for Our Beginning: An Enquiry, Based on Meteorite Research, into the Origin of Our Planet and Life.* Oxford, England: Oxford University Press, 1983. A well-written, well-illustrated introduction to meteorites and their relationship to planetary formation. Although the book is technical, it is understandable for the average reader. Best suited to college-level readers.

Mason, Brian. *Meteorites.* New York: John Wiley & Sons, 1962. This is perhaps the best of the early books on meteorites and their importance to science. Although the book is dated in the light of modern technology, it remains an excellent primer for the study of meteorites.

McSween, Harry Y., Jr. *Meteorites and Their Parent Planets.* 2d ed. New York: Cambridge University Press, 1999. An overview of the study of extraterrestrial debris. Good reference for scientists, students, and amateur astronomers.

Norton, O. Richard. *The Cambridge Encyclopedia of Meteorites.* New York: Cambridge University Press, 2002. Thoroughly discusses interior, external, and atomic compositions of meteorites. Valuable resource for scientists, students, and meteorite enthusiasts.

_____. *Rocks from Space: Meteorites and Meteorite Hunters.* 2d ed. Missoula, Mont.: Mountain Press, 1998. Introduces the reader to meteorites, asteroids, comets, and impact craters. This revised and updated edition includes dozens of new photographs.

Norton, O. Richard, and Lawrence Chitwood. *Field Guide to Meteors and Meteorites.* London: Springer, 2008. A guide to aid readers in observing meteors as well as locating and analyzing meteorites.

Reynolds, Mike. *Falling Stars: A Guide to Meteors and Meteorites.* Mechanicsburg, Pa.: Stackpole Books, 2001. A guide for nonscientists to observing meteors and meteorites, photographing them, and recording data about them.

Sears, D. W. *The Nature and Origin of Meteorites.* Bristol, England: Adam Hilger, 1978. A very readable introduction to the study of meteorites, especially in its historical treatment and its review of the basic concepts. The book does go quite deeply into specialized areas, but it will help the casual reader gain a better perspective on the subject matter. Best suited for the college level.

Spangenburg, Ray, and Kit Moser. *Meteors, Meteorites, and Meteoroids*. Secaucus, N.J.: Franklin Watts, 2002. Information about meteorites, meteors, and meteoroids. This book also discusses impacts with Earth, watching and hunting meteorites, and what can be learned about the universe from them.

Wasson, John T. *Meteorites: Their Record of Early Solar-System History*. New York: W. H. Freeman, 1985. This book is a well-written and well-illustrated introduction to the science of meteoritics. The author covers most of the significant topics in a clear and understandable way and offers a wealth of information for both the casual reader and the serious student.

METEORITES: STONY IRONS

Category: Small Bodies

Stony-iron meteorites are intermediate in composition between stony meteorites and iron meteorites. The two major types of stony-iron meteorites are the pallasites and the mesosiderites. The study of pallasites provides evidence for constraints on planetary differentiation processes. The mesosiderites record a history of repeated impacts of projectiles on the basaltic surfaces of their parent body.

Overview

Meteorites are divided into three broad categories: stony meteorites (or stones), nickel-iron meteorites (or irons), and a group called the stony irons that have both stone and iron components. These stony-iron meteorites are quite rare, constituting only about 1 percent of all the meteorites recovered soon after their fall to Earth was actually observed. The stony irons are more important than their low abundance suggests, however, since they provide a link between the stones and the irons and serve as probes of certain planetary processes. There are four distinct types of stony-iron meteorite: pallasites, mesosiderites, siderophyres, and lodranites. The pallasites and mesosiderites are the most common stony irons; the siderophyres and lodranites are quite rare, represented by only a few specimens each.

Pallasite meteorites are composed of millimeter- to centimeter-sized angular or rounded fragments of magnesium-rich olivine set in a continuous matrix of nickel-iron. In these meteorites, the olivine content ranges from 37 to 85 percent by volume, with the nickel-iron metal accounting for almost all of the remaining material. The minerals troilite, schreibersite, and chromite are sometimes found in small amounts.

The detailed process by which the pallasites formed is still a subject of scientific debate, but they appear to sample a boundary region where nickel-rich iron was in contact with silicate crystals, an environment analogous to the Earth's core-mantle boundary. One mechanism for the formation of pallasites could have been the heating and consequent differentiation of a chondritic parent body. The high-density iron-nickel-sulfur liquid settled to the center, forming a molten core and leaving a silicate-rich mantle. As the mantle cooled, olivine, which is generally the first silicate mineral to crystallize out of cooling silicate liquids of a wide range of compositions, formed and settled to the core-mantle boundary.

The mechanism by which molten metal from the core surrounded the olivine crystals to produce the pallasite structure is not yet understood. It has been proposed that perhaps the mantle shrank as it cooled, squeezing molten metal out of the core and into the olivine-rich layer. Alternatively, the core may have contracted during cooling, causing the olivine layer to collapse into the void, giving rise to the mixing. Further cooling would have resulted in the solid pallasite material, which later was excavated from the parent body by major impacts. Therefore, the pallasites are thought to provide samples similar to the core-mantle boundary region on Earth.

Comparison of the chemistry of the pallasites with that of the Earth provides some constraints on the Earth's formation and differentiation process. The Earth is generally assumed to have formed with the same chondritic composition as the pallasite parent body. After differentiation, the concentration of nickel in the Earth's upper mantle remained at about 0.2 percent. The silicates in the pallasites are much more depleted in nickel, having a concentration of only 0.002 percent. One possible explanation for the additional nickel in the Earth's outer layers is that after differentiation, additional chondritic material was added to the surface, presumably by impacting objects.

Constraints on the size of the pallasite parent body come from a study of how fast these objects cooled after differentiation. If two objects start at the same temperature and are allowed to cool, the smaller object will cool more rapidly, since it has a larger ratio of surface area to volume than does the larger object. The cooling rates determined for the pallasites, and the iron meteorites related to them, are consistent with formation in an object much smaller than the Earth's moon, perhaps no larger than 10 kilometers in diameter. The texture, composition, and cooling rate of typical pallasites are consistent with their metal being related to a group of iron meteorites called the IIIAB irons. If so, then samples of the pure core material of the pallasite parent body are also available as the IIIAB irons.

The differentiation process believed to have occurred in the early history of the Earth and of the pallasite parent body has been simulated in the laboratory by heating chondritic meteorites. As the temperature increases, the meteorites melt in stages. The first liquid to appear is composed mainly of iron, nickel, sulfur, and trace elements that have an affinity for these major elements. Because this liquid is twice as dense as the remaining silicates, it sinks to the bottom. Further melting yields liquids of basaltic composition and a solid residue of mostly olivine. The basaltic liquid, which is less dense than the solids, floats to the top. When cooled, the resulting structure has metal at the bottom, an olivine layer in the middle, and basaltic material on top. For the Earth, this process would give rise to a dense metal core surrounded by an olivine-rich mantle and covered with a basaltic crust. The absence of samples from the Earth's deep interior, however, prevents direct verification of this structure.

Examination of the pallasite meteorites strongly suggests that the pallasite parent body formed with a chondritic composition, was heated and melted, differentiated into a metallic core and silicate mantle, and subsequently cooled and solidified. Thus, the pallasites confirm that planetary differentiation took place on the pallasite parent body in the same manner as proposed for the Earth.

The mesosiderite meteorites are quite different from the pallasites. They are composed of angular chunks of basaltic rocks and rounded masses of metal. The metal phases constitute 17 to 80 percent of the mesosiderites by weight. The major silicate minerals are plagioclase feldspar, calcium-rich pyroxene, and olivine. The mesosiderites are polymict breccias; that is, they are composed of fragments of unrelated rocks. They contain pyroxene-rich fragments, like the diogenite achondrite meteorites, and fine-grained fragments of eucrite achondrite meteorites. The eucrites and diogenites are composed of magmatic rocks similar, respectively, to terrestrial basalts and cumulates.

The mesosiderites appear to have formed from repeated impacts on an asteroidal surface, which brought together at least three distinct types of material: diogenitic and eucritic rocks from the surface of the asteroid and a nonindigenous metallic component, possibly from the impacting objects. If the metal fragments in the mesosiderites are projectile material from the core of a previously fragmented asteroid, these fragments must have struck the surface of the diogenite-eucrite parent body at a very low velocity. Impacts at velocities higher than about 1 kilometer per second lead to very low concentrations of the projectile material in the resulting breccias. This low-impact velocity would suggest that the parent body exerted a very small gravitational attraction on the falling metal, indicating that the diogenite-eucrite parent was a relatively small asteroid, not a planet-sized object.

The mesosiderites are similar to lunar surface breccias, which also formed by multiple impacts into basaltic rock. They allow the processes of basaltic volcanism and impact brecciation to be examined in a different solar system region and an earlier time than occurred on the Moon.

There are a few other meteorites that contain mixtures of metal and silicate phases, but they are otherwise dissimilar to the pallasites and mesosiderites. The siderophyre type, represented only by the single meteorite known as Steinback, is composed of the silicate mineral bronzite (an iron-magnesium pyroxene) and metal. The lodranites are composed of olivine, calcium-poor pyroxene, and metal. These two rare types of stony-iron meteorite have not been as well studied as the pallasites and mesosiderites.

Methods of Study

The pallasite meteorites have been well studied by a variety of techniques, because, along with the iron meteorites, they provide a window on the processes and conditions in the deep interior of their parent bodies and clues to similar processes thought to have occurred on Earth. Much of the evidence concerning these processes comes from detailed analyses of the chemical abundances of major and trace elements in individual minerals from each meteorite. Detailed modeling of the differentiation process suggests that certain metal-seeking trace elements will concentrate in the metallic core, while other trace elements concentrate in the silicate mantle.

Early studies of the metal phases in iron and stony-iron meteorites were done by examining their textures, because the abundances of the trace elements were difficult to determine. More recently, however, the abundances of these trace elements, present at the level of no more than a few atoms in every million atoms of bulk material, have been measured by neutron activation, X-ray fluorescence, and electron microprobe analysis.

Detailed measurements of the abundance of trace elements, emphasizing the elements gallium, germanium, and iridium, in iron meteorites, show twelve to sixteen distinct compositional clusters, indicating that the irons sample a minimum of twelve different parent bodies. Almost all the pallasites have metal compositions and textures, suggesting that they are related to a single group of iron meteorites, the IIIAB irons. This relationship suggests that the metal portion of the pallasites samples the core of the same parent body as the IIIAB irons.

Chemical analysis of the olivine grains in pallasites indicates that the olivine has a very narrow range of compositions. Within each meteorite, the olivine crystals are homogeneous; that is, they show no significant compositional variation from grain to grain. This narrow range of olivine compositions suggests that the grains formed from a silicate liquid of uniform composition. Most of the pallasites, then, appear to sample the core-mantle boundary of a single parent body.

A few pallasites differ from the majority in that they contain olivine richer in iron than the common pallasites. A few of these pallasites are also enriched in nickel and the trace elements germanium and iridium and depleted in gallium relative to the common pallasites. The trace element abundances suggest that these pallasites sample a parent body different from the common pallasites; however, the metal in these cannot be identified with any iron meteorite group.

The cooling rates of pallasite meteorites, from which the size of the parent bodies can be inferred, are determined by examination of the metal. Meteoritic metal consists of two distinct nickel-iron alloys: kamacite, an alloy that can be no more than 7 percent nickel, and taenite, which frequently has more than 20 percent nickel. Metallic liquid cores generally are thought to have a higher nickel content than can be accommodated in the kamacite structure alone. As the metal cools, the amount of kamacite increases, and nickel atoms diffuse from the newly formed kamacite into the nearby taenite. Since nickel diffuses more rapidly in kamacite than in taenite, however, the nickel will build up at the kamacite-taenite boundaries. There will therefore be more nickel at the edges of the taenite than near the center. This distribution of nickel in the taenite varies with the cooling rate.

Electron microprobe analysis of the taenite grains gives the abundance of nickel as a function of distance from the edge, which allows the cooling rate to be estimated. The nickel distribution in the metal of the normal pallasites is consistent with a very rapid cooling rate, implying an extremely small parent body (less than 10 kilometers in diameter). The same technique suggests that the cooling rate of the IIIAB iron meteorites, with which the pallasites are apparently associated, was somewhat slower, implying a parent body of 200 to 300 kilometers in diameter. The reason for this difference is not yet understood.

The mesosiderites, although they are also mixtures of stone and metal, are quite different from the pallasites. The silicate portion of the mesosiderites is rich in the minerals plagioclase feldspar and calcium-rich pyroxene. These minerals melt at relatively low temperatures and are common on the surface of the Earth and the Moon. Unlike the olivine found in the pallasites, the basaltic minerals found in the mesosiderites were probably never in direct contact with the metal in the core of the parent body.

Detailed examination of the mineralogy of the mesosiderites shows that they consist of a mixture of three distinct components, each represented by a distinct type of meteorite. The silicates are fragments of both eucrite and diogenite achondrite stony meteorites; the metal phases resemble the iron meteorites. The eucrites are basaltlike meteorites composed mainly of calcium-poor pyroxene and plagioclase feldspar thought to have crystallized on or near the surface of their parent body. The diogenites are composed mainly of bronzite (an iron-magnesium pyroxene) which resembles the pyroxene cumulates found in the Stillwater complex and in other layered terrestrial intrusions. This combination of eucrite and diogenite fragments in the stony-iron mesosiderites, and in the howardite achondrite stony meteorites, is taken as evidence that the eucrites and diogenites formed on the same parent body. The metal in the mesosiderites occurs mainly in the form of nuggets, sometimes up to 9 centimeters in diameter, or fragments. They are quite distinct from the metal veins that are continuous throughout the pallasites. Trace element analysis of the metal in the mesosiderites suggests a similarity with the IIE iron meteorites; however, the link is much weaker than that of the common pallasites with the IIIAB iron meteorites.

Context

The process of differentiation (the melting of a primitive parent body and the concentration of iron-nickel-sulfur in the core and the lighter silicate minerals in the mantle) is believed to have taken place on Earth. The study of the pallasite meteorites, which are composed of a mixture of iron-nickel core material and olivine mantle material, confirms that this process of differentiation did occur early in solar-system history, at least on the pallasite parent body. Though the chemical compositions of most pallasites are consistent with a single parent body, the few pallasites of unusual composition suggest that the pallasite meteorites sample core-mantle boundaries of several parent bodies. That indicates that the differentiation process was relatively common. The good match between the postulated chemical compositions of the core and mantle and the compositions actually seen in the metal and olivine phases of the pallasites confirms the model of chemical segregation developed for the differentiation of the Earth. It is believed, however, that the common pallasites sample the core-mantle boundary of a much smaller body than the Earth.

The mesosiderites, although they also consist of a mixture of metal and silicates, are quite different from the pallasites. The silicates in the mesosiderites apparently sample basaltic material similar to that found in the lunar surface, in the Earth's crust, and, as implied by chemical measurements taken by spacecraft, on the surfaces of Mars and Venus. The iron fragments may be projectiles that struck the rocky surface of the mesosiderite parent body and were incorporated into the rock produced by the impact. The mesosiderites demonstrate that basaltic rocks similar to those on Earth occur on the surfaces of some asteroids and that the impact processes that dominate the lunar landscape also occurred in the early history of the solar system.

George J. Flynn

Further Reading

Bevan, Alex, and John De Laeter. *Meteorites: A Journey Through Space and Time*. Washington, D.C.: Smithsonian Institution Press, 2002. A pictorial work that also deals with radioactive dating and geologic composition of meteorites. Good for general audiences.

Dodd, Robert T. *Thunderstones and Shooting Stars: The Meaning of Meteorites*. Cambridge, Mass.: Harvard University Press, 1986. This reference book explains why there is a scientific interest in meteorites and summarizes what is known about them. It includes chapters on why planets melt and on the iron and pallasite meteorites. Although written as a college-level text, it provides detailed explanations of the phenomena and techniques of analysis that will be understood by the reader without an Earth science background.

Kerridge, John F., and Mildred S. Matthews, eds. *Meteorites and the Early Solar System*. Tucson: University of Arizona Press, 1988. A collection of articles by sixty-nine contributing authors describing the state of meteorite research as it relates to early solar system processes. Section 3.2 focuses on igneous activity and the process of differentiation on the parent bodies of the iron and stony-iron meteorites. Aimed at graduate students studying meteoritics, with a detailed review and comprehensive bibliography of the major topics in meteorite research. Well illustrated.

McSween, Harry Y., Jr. *Meteorites and Their Parent Planets*. 2d ed. New York: Cambridge University Press, 1999. An overview of the study of extraterrestrial debris. Good reference for scientists, students, and amateur astronomers.

Norton, O. Richard. *The Cambridge Encyclopedia of Meteorites*. New York: Cambridge University Press, 2002. Thoroughly discusses interior, external, and atomic compositions of meteorites. Valuable resource for scientists, students, and meteorite enthusiasts.

_____. *Rocks from Space: Meteorites and Meteorite Hunters*. 2d ed. Missoula, Mont.: Mountain Press, 1998. Introduces the reader to meteorites, asteroids, comets, and impact craters. This revised and updated edition includes dozens of new photographs.

Norton, O. Richard, and Lawrence Chitwood. *Field Guide to Meteors and Meteorites*. London: Springer, 2008. This guide is designed to aid readers in observing meteors, as well as locating and analyzing meteorites.

Reynolds, Mike. *Falling Stars: A Guide to Meteors and Meteorites*. Mechanicsburg, Pa.: Stackpole Books, 2001. A guide to observing meteors and meteorites, photographing them, and recording data. For non-scientists.

Spangenburg, Ray, and Kit Moser. *Meteors, Meteorites, and Meteoroids*. Secaucus, N.J.: Franklin Watts, 2002. Covers impacts with Earth, watching and hunting meteorites, and what can be learned about the universe from them.

METEOROIDS FROM THE MOON AND MARS

Categories: Mars; Small Bodies

Prior to the 1969 Apollo lunar landings, meteorites represented the only extraterrestrial material available for scientific study. In 1979 scientists found an unusual meteorite in Antarctica that resembled a certain type of moon rock. Numerous comparative studies confirmed that this Antarctic meteorite was of lunar origin and the result of an impact event. Subsequent studies on a small group of anomalous meteorites provided evidence to support their Martian origin.

Overview

Throughout human history people have been aware of rocks falling from the sky. In earlier times they were thought to be either gifts from the gods or the result of severe weather picking up rocks from one location and then dropping them at another. Later, as human curiosity developed into science, people began to look differently at these strange rocks and associated the appearance of a bright fireball with the fall of a meteorite. They also noted that meteorites were composed of different substances. Some were made of nearly pure metal, while others had the appearance of some volcanic rocks. Scientists later performed chemical and mineralogical analyses that identified elements and minerals in meteorites that were either rare or unknown on Earth. Eventually all this evidence led to the conclusion that meteorites originated in space and were not from the Earth.

Meteorites were once the only extraterrestrial material available for scientific study. That changed when the six Apollo lunar landings returned to Earth more than 383 kilograms of rocks and other lunar samples. Later, Soviet Luna missions returned a small sampling of lunar material. Among the rock specimens collected were examples of breccia, basalt, anorthosite, and gabbro. Along with regolith samples, this sampling of lunar rocks has provided scientists with enough information to gain a good understanding of the Moon's composition, internal structure, and geological history throughout time. However, Apollo samples were limited to only six locations, and a couple of these sites were very similar to each other in their geological settings. Vast areas of the lunar surface, including the lunar far side, were not represented. Scientists were anxious to get their hands on rock specimens from these unexplored regions. Little did they know

that meteorites would once again provide the required research material.

Along with the 1969 Apollo lunar landings, another rather important event took place in planetary science that year. In Antarctica, a team of Japanese scientists working in the Yamato Mountains region came across nine meteorites sitting on the ice in close proximity to one another. Later, after these meteorites had been classified, it was determined that they represented several different falls. This was extraordinary and suggested that Antarctica might be a "treasure house" for meteorite finds. Subsequent search teams have recovered tens of thousands of meteorites, confirming the theory that Antarctica offers a unique situation for meteorite finds. Found among this huge number of specimens is the full range of stone, stony-iron, and iron meteorite types. Also present are a relatively large number of the biologically important carbonaceous chondrites and a variety of achondrites.

During the 1979-1980 Antarctic field season, another group of Japanese scientists searching for meteorites in the Yamato Mountains recovered what was at first believed to be an anomalous achondrite. Upon later examination, researchers recognized the appearance of a lunar breccia. After additional analyses and comparison to Apollo lunar rocks, this specimen was determined to be a lunar meteorite. It had been blasted off the Moon's surface during a large impact event and was later pulled to Earth and entered the atmosphere just like any other meteorite. Confirming that this specimen came from the Moon could not have been possible without the availability of the Apollo lunar specimens for comparison.

Confirmation that lunar rocks have reached the surface of Earth as meteorites excited the scientific community. Not only did these finds provide additional lunar material for study; they also represented material from areas on the Moon that had not been explored by the Apollo astronauts. Each year, as more lunar meteorites were being found in Antarctica, meteorite dealers and collectors from around the world began to search other locations to find lunar meteorites. Soon the deserts of Australia, Northwest Africa, Libya, and Oman became prime search areas for meteorites. Within a relatively short period, meteorites, found by the local inhabitants, began turning up in marketplaces and were quickly purchased by mineral dealers. Gradually these meteorites found their way to the scientists for study. Found among the huge numbers of ordinary meteorites were the occasional gems, the lunar meteorites and the equally exciting SNC meteorites.

Found in Antarctica in 1981, this meteorite—which includes white fragments of anorthite, typical of the lunar highlands—was among the first to show that not all meteorites originate in the asteroid belt. (NASA/Johnson Space Center)

Prior to the discovery of lunar meteorites in Antarctica, a small group of achondrite meteorites called SNCs (pronounced "snick" and standing for shergottites, nakhlites, and chassignites), could not be readily explained. They were different from most meteorites because of their relatively young radiometric ages and the fact that they were essentially volcanic rocks. When compared to the isotopic chemistry of Earth or Moon rocks, they were clearly different and their origin was unknown. At this time it was generally believed that all meteorites came from the asteroid belt as a result of numerous impact events early in the history of the solar system. This clearly was not the case for the SNCs. As early as 1979, based on the new understanding of lunar meteorites, some scientists suggested that the SNCs might have come from Mars. Initially most scientists did not readily accept this theory, but that quickly changed. A couple of the suspected Martian meteorites contained glass in which a tiny amount of gas had been trapped long ago, while the rock was on Mars.

When scientists analyzed this gas it was found to have a nitrogen and noble gas content that closely matched the gases found in the Martian atmosphere, as sampled by the Viking Landers in 1976. Also, many geologists believed that the massive volcanoes on Mars such as Olympus Mons could have been active at the time that these meteorites were believed to be on Mars. When scientists connected the relatively young age of the SNC meteorites to the volcanoes, it seemed to make sense that the SNCs might be of Martian origin.

To many scientists it seemed quite clear that the best place of origin for these intriguing SNC meteorites had to be Mars. For others, the evidence was not as convincing. The main argument against a Martian origin for the SNC meteorites rests on the fact that scientists do not have a documented sample of a Martian rock for comparison. No astronauts or robotic explorers have collected samples from the Martian surface and returned them to Earth. With no definitive Martian rocks for comparison, the

SNCs cannot be verified as truly Martian. The robot explorers Pathfinder, Spirit, and Opportunity have traveled across Mars and viewed hundreds of rocks close up, but they have not seen a type of rock that matches any of the SNC meteorites. Although this lack of evidence does not preclude the possibility that SNCs are of Martian origin, it will take a major sampling effort by both robotic and human explorers to make that determination, and therefore the debate over the origin of the SNC meteorites will continue to spark scientific discussions for years if not decades to come.

Knowledge Gained

The basic study of meteorites paved the way for the development of the scientific technology that was required for the examination of materials brought back from the Moon by the Apollo astronauts. Geologists and geochemists studying meteorites have a more difficult task than those who study Earth rocks. The average meteorite fall usually consists of a relatively small amount of material. Proper classification and additional detailed studies require destructive analyses. With only a small amount of material available, scientists had to develop analytical techniques that require only minimal amounts of the precious material available to them. Instruments such as the electron microscope, electron and ion microprobes, and mass spectrometer gave the researcher tools to explore the secrets of the universe from the chemical composition of a meteorite. As is usually the case, technology specifically developed for pure scientific research was immediately integrated into the commercial market and became valuable tools for industry and medical applications.

The rock and regolith samples returned by the six Apollo landings were not representative of the entire Moon and its major geological terrains. Many questions concerning the Moon's origin and relationship to Earth could not be answered with only the Apollo material. Scientists would have to wait patiently for some future missions to bring back samples from different areas. The 1979 find of lunar meteorites in Antarctica changed all that, giving scientists the opportunity to study additional lunar materials. A lunar meteorite represents material blasted off the Moon's surface during an impact event; hence, it does not represent a single location but could have originated anywhere on the lunar surface. By examining lunar meteorites that originated from different locations on the Moon, scientists have been able to construct a more comprehensive picture of lunar geology.

The case for Martian meteorites proved to be equally exciting, especially with the meteorite ALH 84001. This unique meteorite was found during the 1984-1985 field season in the Allan Hills region of Antarctica. Initially classified as an anomalous achondrite, it was later reclassified as a SNC Martian meteorite. Then in 1996, after years of intensive study, scientists at the National Aeronautics and Space Administration (NASA) announced the discovery of possible microscopic fossil bacteria within carbonate deposits present in the meteorite. This revelation created much excitement within the scientific community and sparked intense controversy and debate. Additional studies provided evidence that the features in the meteorite resembling bacteria could have been the result of an inorganic process, too. Neither side in the debate could definitely prove the other wrong, so in the end the "fossil" was deemed inclusive with regard to the evidence of organic material, and scientists concluded that additional evidence was required.

Although the argument for fossil bacteria in ALH 84001 was not fully accepted, it inspired other scientists to look for microscopic evidence of life in more promising meteorites, notably the carbonaceous chondrites. The recognition of lunar and Martian meteorites also suggests that planets can exchange material with each other and possibly distribute the essential chemical compounds of life through giant impacts.

Context

Confirmation of lunar meteorites led to the possibility of meteorites coming from Mars. Once scientists became comfortable with the idea that rocks could be blasted off the Moon's surface and pulled to Earth, they expanded their options to include other worlds. Because of its relative closeness to Earth and similar geological features, Mars became the logical first choice. Using the same technology developed for lunar studies, scientists began to look at a number of intriguing anomalous meteorites (SNCs) with uncertain places of origin. By comparing gas compositions and isotopic data from these meteorites to data obtained through the Viking Mars Project, it was concluded that they could have come from Mars in a manner similar to the way lunar meteorites arrived. Although this Mars-origin theory for these meteorites is not conclusive, most scientists seem to have accepted it as fact and moved on from there. Now other anomalous types of meteorite are being examined for evidence that may have them coming from Mercury or possibly Venus. Perhaps one day a particularly interesting meteorite may produce an ancient piece of Earth rock tossed out by an early giant impact event that has finally returned home.

Paul P. Sipiera

Further Reading

Bevan, Alex, and John De Laeter. *Meteorites: A Journey Through Space and Time*. Washington, D.C.: Smithsonian Institution Press, 2002. A beautifully illustrated book that covers the topic of meteorites at a level that is suitable for a wide readership.

Hartmann, William K. *Moons and Planets*. 5th ed. Belmont, Calif.: Thomson Brooks/Cole, 2005. This book deals principally with the geological aspects of the entire solar system as related to planetary formation. An excellent resource for students of astronomy and geology.

Kring, David A. "Unlocking the Solar System's Past." *Astronomy* 34, no. 8 (August, 2006): 32-37. A good introduction to the subject of meteorites and how they relate to the other members of the solar system. Suitable for general readers.

Lauretta, Dante S., and Harry Y. McSween, eds. *Meteorites and the Early Solar System II*. Tucson: University of Arizona Press, 2006. A comprehensive collection of scientific papers covering the role meteorites play in planetary formation. This book serves a valuable resource for the graduate student and the professional researcher alike.

Norton, O. Richard. *The Cambridge Encyclopedia of Meteorites*. Cambridge, England: Cambridge University Press, 2002. A comprehensive and well-illustrated work on the topic of meteorites. Suited for a wide audience, from high school through graduate students and meteorite enthusiasts.

METEORS AND METEOR SHOWERS

Category: Small Bodies

Meteors are those streaks of light produced by small solar-system bodies (meteoroids) entering the Earth's atmosphere. Fragments from asteroids produce sporadic meteors, while debris left along the orbit of a comet causes meteor showers. Both provide information about the origins of the solar system, especially if they reach the ground and are recovered as meteorites.

Overview

Scientific study of meteors and their relation to meteorites did not start until the beginning of the nineteenth century. Earlier meteorite falls were observed, with stones recovered, but most witnesses were ridiculed, and "sky stones" were treated with suspicion. In the Bible, Joshua 10:11 records a battle in which the enemy was defeated by "stones from heaven," which may have been meteorites. Acts 20:35 refers to the image of Diana of Ephesus standing on a stone that fell from heaven.

Anaxagoras, Plutarch, and several Chinese recorders from as early as 644 B.C.E. described stones falling from the sky. A stone preserved in a corner of the Kaaba in Mecca fell in the seventh century. The oldest authenticated meteorite in Europe, a 120-kilogram stone that fell in Switzerland in 1492, is still preserved in a museum. In spite of this evidence, much doubt remained among scientists in Europe. When a stone fell near Luce in France in 1768, it was studied by French chemist Antoine Lavoisier and two other French scientists, who concluded that it was an ordinary stone struck by lightning.

In 1794, Czech acoustic scientist Ernst Chladni published an account of numerous reported meteorite falls, giving strong evidence that some of them must be of extraterrestrial origin. He stated that the flight of such an object through the atmosphere caused the bright, luminous phenomenon known as a fireball. Chladni found few supporters for this idea, as most held to Aristotle's view that comets and flashes of light across the sky were atmospheric phenomena (the word "meteor" comes from the Greek word for things related to the atmosphere, as in meteorology). Chladni's cosmic theory of meteors was finally confirmed in 1798 by two students at the University of Göttingen, H. W. Brandes and J. F. Benzenberg, who had read his book. They made simultaneous observations of "shooting stars" from two different locations separated by several kilometers and used a simple triangulation method to show that the light flashes originated at least 80 kilometers above the ground from objects moving several kilometers per second from a source beyond the atmosphere. Most doubts about meteorite falls were removed after the physicist Jean-Baptiste Biot reported an unusual fall of two or three thousand stones at L'Aigle in 1803, which eyewitnesses said was preceded by a rapidly moving fireball and explosion.

On a clear, dark night, a diligent observer may be able to witness on average six meteors per hour. More are visible after midnight than before, increasing to a maximum just before dawn. In the 1860's, Italian astronomer Giovanni Schiaparelli, famous for his discovery of Martian *canali* (channels), explained the increase in meteors at certain times as resulting from the Earth's orbital and rotational motion. Before midnight,

The Leonid meteor shower, in an image captured by an aircraft on November 17, 1999. The Leonids occur when Earth passes through the debris field of Comet Tempel-Tuttle. (NASA/SAS/Shinsuke Abe and Hajime Yano)

the observer is on the trailing side of the Earth's motion and can see only those meteors that overtake the Earth. After midnight, an observer is on the leading side of the Earth's motion and will intercept meteors in front of it. Thus, meteors will appear brighter because they are entering the Earth's atmosphere at a higher velocity. Because the Earth's orbital velocity is about 30 kilometers per second, and the escape velocity from the Sun at the Earth's orbital distance is about 43 kilometers per second, solar-system objects should range in speed from 13 to 73 kilometers per second. Because no meteors have been observed with a faster speed, it is believed that they come from within the solar system rather than from interstellar space.

Most meteors become visible about 100 kilometers above the Earth's surface and are completely consumed when they reach about 70 kilometers, although a few larger ones reach about 50 kilometers. Most meteors range in size from a few microns up to several millimeters. Survey estimates indicate that about 25 million meteors are bright enough to be seen over the entire Earth in any

twenty-four-hour period. Telescopic surveys suggest that several billion meteoroids enter the Earth's atmosphere every twenty-four hours, with an average total mass of about 100,000 kilograms. Most of this is consumed in the atmosphere as meteoroids are heated by friction to incandescence, but on average about 1,000 kilograms per day are deposited on the Earth as meteorites.

More than half of all meteors are called sporadic because they appear at any time and from any direction in the sky. The remaining meteors are associated with meteor showers that appear to radiate from a common point in the sky, called the radiant; they actually move along parallel paths but appear to diverge from the radiant—much like the divergence of railroad tracks when viewed in perspective. Meteor showers recur on an annual basis, with about a tenfold increase over the usual sporadic rate. They are named for the constellation in which the radiant appears to be located. Annual showers occur when the Earth crosses a meteoroid stream that fills the orbit of a comet, while periodic showers occur less frequently, when Earth crosses a meteroid swarm in the wake of a comet. The most spectacular periodic meteor showers are the Leonids, whose radiant is located in the constellation Leo. Historical records as far back as 902 C.E. mention the Leonids. A spectacular display on October 14, 934, is described in Chinese, European, and Arabic chronicles. The Japanese recorded a six-hour display in 967, and Chinese records continued to describe them every thirty-three years for several centuries.

The modern study of meteor showers began with the famous naturalist Baron Alexander von Humboldt. He observed the Leonids by chance during a trip to South America in 1799 in a two-hour display of hundreds of thousands of meteors. Humboldt was the first to suggest that these meteors might originate from a common point in the sky. The greatest Leonid display in the nineteenth century was observed in the United States and Canada on November 12, 1833. About one thousand meteors per minute were counted, and the appearance of the radiant was confirmed. The following year, two Americans, D. Olmstead and A. C. Twining, suggested that the annual Leonids were caused by the Earth passing through a cloud of meteoroids each November. A few years later, German astronomer Heinrich Wilhelm Olbers proposed that the more intense periodic meteor showers of 1799 and 1833 were caused by a denser swarm of the Leonid meteoroid stream. In 1864, H. A. Newton of Yale College reached the same conclusion independently and showed a period of recurrence of just over thirty-three years from historical records, beginning with the shower of 902. Their prediction of a spectacular display in 1866 was confirmed. Later, English astronomer John Couch Adams, who theorized the existence of the planet Neptune, succeeded in computing the Leonid stream orbit.

In the 1860's, other meteoroid streams were identified and traced back through history. Records back to the tenth century in England recorded meteor showers associated with the festival of St. Lawrence (August 10), known as "the tears of St. Lawrence" but now identified as the August Perseids from their radiant in Perseus. In 1861, the American astronomer Daniel Kirkwood, who later discovered gaps in the asteroid belt, suggested that meteor showers result from debris left in the wake of a comet through which the Earth occasionally passes. In 1866, Schiaparelli announced that the August Perseids appear to occupy the same orbit as Comet Swift-Tuttle (1862 III). Soon after, Urbain Le Verrier and C. A. F. Peters identified the November Leonids with Comet Tempel-Tuttle (1866 I), which had a recurrence period of thirty-three years.

Both the May Aquarids and the October Orionids have been associated with Halley's comet. The greatest naked-eye meteor observer was W. F. Denning, who published a catalog in 1899 of several thousand radiants, mostly of minor meteor showers of less than 10 meteors per hour, based on more than twenty years of observation.

Like comets, meteor streams may be perturbed by planets into new orbits. Those with high inclinations to the ecliptic plane (the plane of the Earth's orbit) or in retrograde orbits (opposite to the Earth's motion) are least affected, such as the Leonids, Perseids, and Lyrids. After the Leonid display of 1866, the main body of the stream passed close to Jupiter and Saturn. Its associated comet could no longer be found, and only a few meteors were observed in 1899 and 1933. The comet was found again in 1965, and then, on the morning of November 17, 1966, the Leonids returned, with meteors as bright as Venus. Viewed from the western United States, they reached a maximum rate of more than two thousand per minute before dawn, producing the greatest meteor display in recorded history.

Only the bright fireball meteors, sometimes brighter than the full Moon, are produced by meteoroids large enough to survive passage through the Earth's atmosphere and fall to the ground as meteorites. Almost all of these are sporadic meteors; even among the fireballs, less than 1 percent yield meteorites. Dozens of meteorites fall to the surface of the Earth each day, but very few are recovered. About 95 percent of "falls" (seen falling and then recovered) are classified as "stones" (about 75 percent silicates and 25 percent iron), but 65 percent of "finds" (whose associated meteors are not observed) are "irons" (90 percent iron and 8 percent nickel), because irons are easier than stones to identify on the ground as meteorites.

Dozens of craters apparently formed by large meteorites have been identified around the world. The first such identification was made about 1900 by Daniel Barringer at Canyon Diablo in Arizona. This crater is 1.3 kilometers across and 180 meters deep, with a rim rising 45 meters above the surrounding plain. About 25,000 kilograms of iron meteorite fragments have been found in and around the crater. It is estimated that the crater was formed by an explosive impact about 50,000 years ago from a 60-million-kilogram meteorite. In 1908, a brilliant fireball meteor exploded in the Tunguska region of Siberia, leveling trees over a distance of 30 kilometers and killing some 1,500 reindeer. No large crater or meteorite has been found, but its effects were estimated to be equivalent to the explosion of a billion-kilogram meteoroid. In 1972, a fireball meteor with an estimated mass of a million kilograms was photographed in daylight some 60 kilometers above the Grand Teton Mountains before leaving the atmosphere over Canada.

Methods of Study

Information about meteors can be obtained with the unaided eye, but much greater scope and precision results from the use of photographic, radar (radio echo), and space-probe techniques. Modern photographic meteor observations were begun by Fred Whipple in 1936 at the Harvard College Observatory using short-focal-length, wide-angle cameras. These were later replaced by ultrafast Super-Schmidt cameras that could detect meteors as small as a milligram. To measure the height, direction, and velocity of a meteor, simultaneous photographs of the meteor trail are taken from two stations separated by about 50 kilometers. Each photograph shows the positions of the meteor trail against the background of stars from each station so that its trajectory can be calculated by triangulation. The velocity of the meteor is measured by using a rotating shutter to interrupt the meteor trail up to sixty times per second. The velocity vector and the known position of the Earth in its orbit make it possible to compute meteor orbits.

The density of a meteoroid can be estimated from its deceleration in the atmosphere, showing that most meteoroids are of lower density than are meteorites. Statistical studies have shown that meteoroids with the greatest meteor heights have average relative densities of 0.6, while another group appearing about 10 kilometers lower have relative densities averaging 2.1. The few meteoroids that penetrate deep into the atmosphere have average relative densities of 3.7. Several hundred fireball meteors have been photographed, their masses ranging from 100 grams to 1,000 kilograms, including one meteorite fall near Lost City, Oklahoma. Experiments with artificial meteors and theories of meteor burning led to estimated initial meteoroid masses from observed optical effects. Meteors comparable in light to the brightest stars have initial masses of a few grams and diameters of about 1 centimeter, producing more than a megawatt of power. Meteor showers are produced by the most fragile (lowest-density) meteoroids, and different showers produce meteoroids of different character. In general, short-period comets exposed more often to the Sun produce higher-density particles than do long-period comets, because of greater evaporation.

During World War II, it was accidentally discovered that meteors could be detected by radar. The radar method of studying meteors is especially valuable because it can detect meteors in daylight and is sensitive to meteoroids as small as a microgram. This method depends on the fact that meteors separate electrons from atoms, producing ionized gases that can reflect radio waves. Meteor heights can be measured from the time delay of the return signal, and velocities can be determined from the frequency shift (Doppler effect). Observations from three stations are needed to calculate a meteoroid orbit. Several important meteor showers that occur only in daylight hours were discovered by radar, including the Beta Taurids, which are probably associated with Comet Encke. Radar also shows that radiants can be complex structures that appear to overlap and shift positions within a few hours.

Micrometeoroids with masses of a few micrograms or less have been collected by high-altitude aircraft and rockets. Micrometeoroids are fluffy particles containing carbonaceous material different from normal meteorites but consistent with comet theories. They can be studied with microphone detectors in space probes by measuring the intensity of their collisions. The weak structure of these particles indicates that they are gently separated from their parent material, suggesting dust emitted from evaporating ice in a comet, rather than violently ejected from high-temperature or colliding meteoroids. Particles of less than a milligram contribute the largest fraction of the total mass swept up by the Earth each day. Rocketborne mass spectrometers have recorded metallic ions (charged atoms) of apparent meteoric origin, and meteor spectroscopy has provided chemical analysis of all the major meteor streams. These data indicate significant differences betwen cometary meteor material and the composition of meteorites.

Radioactive-dating techniques indicate that most meteorites have existed as solid bodies for about 4.5-4.7 billion years, close to the estimates for the ages of the Earth, Sun, and Moon. This suggests that all the matter of the solar system condensed at approximately the same time. Cosmic-ray dating from the amount of unusual isotopes produced in a meteorite by cosmic rays colliding with atoms in its crystalline structure usually indicates only a few million years since its formation, presumably by some fragmentation process from a larger asteroid. Fine bands are also observed in such meteorites, similar to those that occur in metal crystals subjected to sharp collisional shock. This finding has led to the idea that meteorites probably come from asteroids that were shattered in collisions.

Context

Meteors and meteor showers not only are interesting as visual phenomena but also provide one of the most important sources of information about asteroid and comet composition and deterioration, as well as clues to the origin of the solar system. Fortunately, most meteors are caused by very small particles (less than 1 gram) and are completely vaporized high in the atmosphere. Meteors enter the Earth's atmosphere with solar-system speeds and random inclinations to the ecliptic (plane of the Earth's orbit); thus, it appears that most meteors are associated with comets or with asteroids that have small inclinations.

A cometary origin for most meteors is supported by the phenomenon of meteor showers, which can be traced to particle swarms in various orbits with random inclinations around the Sun. Many of these showers can be associated with comets or former comets. They appear to be caused by particles released when cometary ices were evaporated by solar radiation. These particles either concentrate in a swarm of meteoroids behind the cometary nucleus or eventually become distributed in a stream around the entire orbit of the comet. Annual meteor showers occur when the Earth crosses a meteoroid stream, while more intense periodic showers occur when the Earth passes through a meteoroid swarm. The densities and compositions of these meteoroids are also consistent with a cometary origin.

The few meteoroids large enough to survive their passage through the atmosphere and yield meteorites are associated with the rare fireball meteors. Their trajectories tend to have low inclinations to the ecliptic, similar to asteroids. The crystalline structure of the metal in meteorites indicates that most were formed at high temperatures and slowly cooled over several million years. Thus, it appears that they did not come from icy comets; rather, they probably originated with asteroids. Calculations show that the rocky outer shell of an asteroid would insulate its hot metallic core, causing it to cool at the very slow rate suggested by the crystalline structure of iron meteorites. Furthermore, the cooling rate in a planet is too slow to fit this observed crystal pattern.

Meteoroids in asteroidal orbits enter the atmosphere with an average velocity of about 20 kilometers per second. Most are slowed rapidly by the atmosphere. If they survive as a meteorite, they simply fall to the ground at free-fall speeds and cool rapidly, since most of the hot surface material is swept off. Meteoroids larger than about one million kilograms (10-meter-sized) strike the ground with most of their initial velocities, producing impact craters. A 50-meter, 100-million-kilogram object moving at high speed can produce a one-kilometer-wide crater, causing widespread devastation by its shock waves and by throwing dust into the upper atmosphere, with marked effects on climate and life on Earth. Some evidence from large craters and geological layers of meteorite debris suggests the possibility that kilometer-sized objects strike the Earth about every twenty-six million years, coinciding with major extinctions of life-forms. One attempt to explain these data theorizes that the Sun has a dim companion star in a twenty-six-million-year eccentric orbit. At its closest approach to the Sun, its gravity would disturb many comets in the outer solar system, causing some of them to strike the Earth.

Joseph L. Spradley

Further Reading

Brown, Peter L. *Comets, Meteorites, and Men*. New York: Taplinger, 1975. This classic book is an excellent and very readable historical study of comets, meteors, and meteorites. Six chapters on meteor showers and meteorites cover the history of these phenomena from earliest times. Appendixes include a table of meteorite impact sites and the major annual meteor showers.

Burke, John G. *Cosmic Debris*. Berkeley: University of California Press, 1986. This book is a scholarly and well-documented history of meteorite and meteor discoveries and theories. Much of the book is suitable for the general reader, with interesting illustrations, but some parts are more detailed and technical. A 50-page bibliography contains about a thousand historical references.

Delsemme, A. H., ed. *Comets, Asteroids, Meteorites*. Toledo, Ohio: University of Toledo Press, 1977. This book is the result of an International Colloquium on the interrelations, evolution, and origins of comets, asteroids, and meteorites. Of the seventy-five articles, twenty-two are on meteors, meteoroids, and meteorites. Although the level is quite technical, the discussion offers a detailed firsthand account of research results for interested students.

Erickson, Jon. *Asteroids, Comets, and Meteorites: Cosmic Invaders of the Earth*. New York: Facts On File, 2003. Part of The Living Earth paperback series, a good basic resource for nonspecialists.

Levy, David H. *David Levy's Guide to Observing Meteor Showers*. Cambridge, England: Cambridge University Press, 2007. A thorough examination of different types of meteorites and a guide to observing meteors and meter showers. For the general reader and the dedicated backyard observer.

_____. *The Quest for Comets: An Explosive Trail of Beauty and Danger*. New York: Plenum Press, 1994. Written by one of the co-discoverers of Comet Shoemaker-Levy 9, this book is for the general reader. It highlights the author's comet discovery program and the comet catastrophe theory.

Morrison, David, and Tobias Owen. *The Planetary System*. 3d ed. San Francisco: Pearson/Addison-Wesley, 2003. Geared for the undergraduate college student, this textbook treats planetary atmospheres as important physical features of the various members of the Sun's family. They are discussed both individually in the context of what is known about each planet's characteristics and with regard to theories about their evolution and the evolution of the entire solar system.

Time-Life Books. *Comets, Asteroids, and Meteorites*. New York: Time Life Education, 1992. Part of the Voyage Through the Universe series. Heavily illustrated and for the general reader.

MOON

Category: Natural Planetary Satellites

The moon is the sole natural satellite of the Earth. Specific astronomical searches have established positively that the Earth has no other satellites larger than a few meters. The lunar body is nearly a sphere with a mean radius of 1738 kilometers (km) or 1000 miles— only 3.7 times less than the Earth. The mean distance of the moon from the Earth is 384,400 km (238,855 miles). The moon is the fifth largest satellite in the solar system and the largest one relative to the size of its planet. The moon is so near and so large in comparison with its "host" that the entire system is often dubbed the "double planet."

Overview

Viewed from above the North Pole of the Earth, the moon travels around it counterclockwise in a slightly elliptical path. The sideric month (one orbit around the Earth with respect to the stars) is 27.3217 days. The synodic month (the cycle of phases visible from the

Earth; for example, the time interval between two successive "new moon phases") is 29.5306 days.

The period of one spin of the moon around its axis (a "lunar day") is exactly equal to the sideric month because of tidal breaking. This phenomenon is also known as "synchronous rotation," or tidal coupling." As a result, from the Earth, people can observe only half of the lunar surface (called the "near," or "visible," "side"). The "far" (called "invisible") hemisphere was photographed for the first time in 1959 by the Soviet robotic spacecraft *Luna-3*, an episode of the space race between the United States and the Soviet Union. On the moon, the disk of the Earth does not rise and set. It is observable only from the near side in an almost permanent point of the lunar sky (fluctuating a little from a small phenomenon called "libration").

The face of the moon was influenced by both internal and external factors. On the surface, observers distinguish so-called darker "maria" (flat "seas" without water) and brighter highlands. All of them are covered with numerous craters, the highlands more so than the seas. The far side of the moon has practically no seas. Because of constant bombardment by various small interplanetary particles, the entire surface is enveloped with thin fractured material called "regolith." There is no atmosphere on the moon. As a result, the difference in temperatures between a lunar day and a lunar night is very high: between –170 degrees Celsius and +130 degrees Celsius (–274 degrees Fahrenheit to +266 degrees Fahrenheit). Water in the form of subsurface ice exists in polar regions. There are no traces of modern tectonics on the surface.

From the Earth, the visible angular diameter of the moon is 0.5 degrees and fairly close to the angular diameter of the sun. This property is essential because sometimes the three bodies, the sun, the Earth and the moon, align along a straight line. In this case, humans observe either a total lunar (if the moon is farther from the sun than the Earth) or a total solar (if the moon is between the sun and the Earth) eclipse. The latter is visible only within narrow strips on the Earth. Such observations are important for solar physics. To see these phenomena, astronomers regularly organize special expeditions. Eclipses often held great religious significance. Scholar Anaxagoras of Clazomenae explained the phenomenon using mathematics. He was imprisoned for asserting that the sun was not a god and that the moon reflected the sun's light.

Astronaut Harrison H. Schmitt standing next to a huge boulder during the Apollo 17 mission to the moon. (National Aeronautics and Space Administration)

The age of the moon is about 4.5 billion years, which is close to the age of the sun and the entire solar system. Of the various concepts of the moon's origin, the prevailing hypothesis is that the Earth-moon system was formed by a giant impact: a planet-sized body hit the nearly formed proto-Earth, ejecting material into orbit around the proto-Earth, which accreted to form the moon.

The mean density of the moon is just 3.34 grams per centimeter^3 and, as a result, the mass of the moon is 81 times less than that of the Earth. The interior of the moon is geochemically differentiated: it has a distinct crust, mantle and core. Surface gravity on the moon is six times less than on the Earth. The general magnetic field of the moon is practically absent.

The moon has always played a significant role in religion, science, art, and culture. Since the Paleolithic, the lunar orb in the sky has been utilized for calendar purposes. That is why the similarity of the terms "moon" and "month" is not coincidental. For the philosopher Aristotle, the moon marked a great border between a mortal and corruptible sublunar (terrestrial) world and an immortal world of ideal heavenly bodies. It became a significant symbol for Islam. For Isaac Newton, the moon was the prime test body to demonstrate mathematically that the fall of an apple and the orbiting of a celestial body are ruled by a single natural law of universal gravity.

Mathematical Modeling

Many mathematicians have developed theoretical models for the motion of the moon. The exact path of the moon around the Earth is affected by many perturbations and is extremely complicated. That is why, after Newton, research of lunar motion (lunar theory) became the central problem of celestial mechanics. Consequently, it appeared among the most critical and difficult tasks for applied mathematics. The moon's gravitational influence on the Earth produces the ocean tides and the tiny lengthening of the calendar year. Most of what we know about the moon's size, shape, and other properties has been derived largely through mathematical computations, using mathematical theory and data from Earth-based observations, satellite imagery, and direct measurements made by astronauts.

Human Exploration

Starting at least from Roman times, science fiction authors were the forerunners for delivering terrestrials to the moon. In reality, the first space robots to the moon were launched by the Soviets in 1959. But they failed in the space race with the United States to realize manned expeditions. The first terrestrials to visit the moon were the American astronauts of the Apollo program. After preliminary robotic programs (Ranger, Lunar Orbiter, and Surveyor) and Apollo flybys, American manned landings on the moon occurred in 1969–1972. Among seven planned landings (from Apollo-11 up to Apollo-17), six missions were tremendously successful. Twelve crewmembers stepped down on the near side of the moon, and six more orbited it. Astronauts performed a number of experiments and returned to the labs about 382 kg of lunar matter. Since 2004, Japan, China, India, the United States, and the European Space Agency have each sent successful automatic lunar orbiters.

Among the many thousands of contributors to lunar programs, mathematicians often played outstanding roles. One significant individual was mathematician Richard Arenstorf, who solved a special case of the three-body problem with figure-eight trajectories now called "Arenstorf periodic orbits." In 1966, he was awarded a NASA medal for exceptional scientific achievement for this work. Another was Evelyn Boyd Granville, who used numerical analysis to aid in the design of missile fuses. She later worked on trajectory and orbit analyses for several space missions, including Apollo. She said, "I can say without a doubt that this was the most interesting job of my lifetime—to be a member of a group responsible for writing computer programs to track the paths of vehicles in space." In fact, mathematicians occupied many seats in the first row of the Mission Control center. Their work was critical for calculating trajectories and for maneuvers that involved the meeting of two objects in space, including landing on the moon. They also played a significant role in determining a rapid and feasible solution that would safely return the damaged Apollo 13 manned spacecraft to Earth.

Among mathematicians in Russia, the most noticeable contribution to flights to the moon was made by Efraim L. Akim of the Keldysh Institute for Applied Mathematics at the Russian Academy of Sciences in Moscow. He was the principal investigator for special lunar orbiters to create a mathematical model of the lunar gravitational field and the leader of a team to calculate trajectories of the Russian lunar robotic spacecraft.

Several international treaties regulate mutual relations of various states with respect to modern space explorations of the moon. The most important among them are the Outer Space Treaty (1967) and the Agreement Governing the Activities of States on the Moon and Other Celestial Bodies (1979).

Alexander A. Gurshtein

Further Reading

Eckart, Peter, ed. *The Lunar Base Handbook.* New York: McGraw-Hill, 1999.

Gass, S. I. "Project Mercury's Man-in-Space Real-Time Computer System: 'You Have a Go, at Least Seven Orbits.'"*Annals of the History of Computing, IEEE* 21, no. 4 (1999).

Stroud, Rick. *The Book of the Moon.* New York: Walker and Co., 2009.

Ulivi, Paolo, and David Harland. *Lunar Exploration: Human Pioneers and Robotic Surveyors.* Chichester, England: Praxis Publishing, 2004.

NEMESIS AND PLANET X

Category: Planets and Planetology

Scientists developed the Nemesis and Planet X theories in an effort to explain why mass extinctions on Earth appear to have occurred roughly every 26 million years. Each theory suggests that a still undiscovered star (Nemesis) or planet (Planet X) periodically perturbs the orbits of comets, sending them into the inner solar system where some could collide with Earth.

Overview

Nemesis and Planet X are theoretical astronomical bodies whose existence has been suggested as a possible driving force for periodic mass-extinction episodes. Nemesis and Planet X theories both propose that about every 26 million years movement of an astronomical body through space causes massive extraterrestrial objects to collide with Earth. The resulting catastrophic changes in the Earth's environment led to the widespread extinction of species. The periodicity of the mass-extinction episodes, and the impacts causing them, are themselves theoretical.

The Nemesis theory hypothesizes that the Sun is orbited by a companion star so small, dim, and distant that astronomers have yet to discover it. This dark, low-mass star is thought to move in a highly elongated orbit at a distance between 25,000 to 150,000 astronomical units (AU) from the Sun. About every 26 million years, when it most closely approaches the Sun, the companion star is believed to pass through the huge "comet reservoir" known as the Oort Cloud. This vast region, which may contain anywhere from 100 billion to 10 trillion widely separated comets, surrounds the solar system, extending to a distance of perhaps 50,000 AU from the Sun. As the companion star moves through the Oort Cloud, its gravitational field disturbs the orbits of nearby comets. Some are deflected into deep space, while others are pushed toward the inner solar system. Of those comets entering the inner solar system,

it is likely that some could collide with Earth. Because of the role that the Sun's companion star is thought to play in causing periodic devastating impacts on Earth, researchers have dubbed it Nemesis, after the Greek goddess of vengeance.

The Planet X theory, which also involves cometary perturbation, proposes that the solar system has a massive, undiscovered tenth planet whose orbit lies beyond that of Pluto. The "X" represents both "the unknown" and the Roman numeral X (ten). Researchers have suggested that this Planet X may have so far escaped detection because its orbit lies outside the plane of the solar system. Planet X is believed to be at least as large as Earth and 50 to 100 astronomical units from the Sun. It is thought to orbit the Sun in about 1,000 years, along a highly elongated, eccentric path. According to this theory, the orbit of Planet X slowly precesses over a period of 58 million years—that is, it regularly changes orientation in space. The precession is such that about every 28 million years (roughly one-half the precession period), Planet X perturbs comets within the Kuiper

Nemesis and Planet X theories both propose that about every 26 million years movement of an astronomical body through space causes massive extraterrestrial objects to collide with Earth (artist's rendition). (Don Davis, NASA)

Belt. This disk of comets located beyond Neptune at a distance of roughly 35 to 1,000 AU from the Sun is regarded by some astronomers as merely the inner portion of the Oort Cloud. Although Planet X has cleared a path through the Kuiper Belt, roughly every 28 million years its precession brings it close to the margins of the cleared area, where comets fall under its gravitational influence. As in the Nemesis scenario, gravitational forces deflect some comets into the inner solar system, leading to periodic cometary impacts on Earth.

Many scientists link such cometary impacts to mass extinctions experienced throughout Earth's history. Earth's fossil record shows that over the past 600 million years there have been several mass-extinction events, forming the boundaries between the major divisions on the geologic timescale. For example, it was in the boundary between the Cretaceous and Tertiary periods (also known, from the German, as the K-T boundary) that the dinosaurs, nearly all marine plankton (tiny photosynthesizing plants), and 15 percent of marine invertebrate families became extinct. Evidence strongly suggests that the K-T extinction 65 million years ago was related to extraterrestrial impact.

If a comet or other large extraterrestrial body were to collide with Earth, the environmental effects would be severe and widespread. The immediate area of impact would be devastated by heat energy and a powerful shock wave; tsunamis and earthquakes might also result. Massive quantities of dust and gases would be thrown into the atmosphere, and a portion of the atmosphere itself might be ripped away. Fireballs formed by coalescing dust and gases would set off large-scale wildfires. The resulting smoke, combined with the dust and gases, would block sunlight for months, reducing global temperatures and impeding photosynthesis.

Depending on where the collision occurred, conditions could be even more severe. If the impact site were rich in limestone and evaporite minerals (as is believed to be the case for the K-T event), sulfur and water vapor released from the rock could combine with atmospheric gases to produce nitric and sulfuric acid, which would eventually fall to Earth as acid rain. In combination with atmospheric gases, carbon freed from the limestone would produce carbon dioxide, a "greenhouse gas" capable of trapping heat and causing global warming. Plant and animal species unable to withstand the extremes and fluctuations in temperature, disruptions in the food chain, and other inhospitable conditions would face extinction.

Methods of Study

The Nemesis and Planet X theories span several disciplines. Paleontologists, geologists, chemists, physicists, and astronomers are among the scientists whose studies have led to the development of these and associated concepts. Both theories are rooted in the idea that extraterrestrial impacts have caused or contributed to extinction events and that these impacts occur periodically.

Some researchers have used iridium data to link extraterrestrial impacts to mass extinctions. Iridium, an element related to platinum, is rare in the Earth's crustal rock (its average abundance is 0.001 part per billion). Gravity drew most of this extremely dense and unreactive element to the Earth's core when the planet was still molten. By contrast, laboratory analysis of meteorites (rocks of extraterrestrial origin found on Earth) has detected

Between the Cretaceous and Tertiary periods, about 65 million years ago, a catastrophic event occurred, which many scientists believe to be an asteroid or cometary impact, that resulted in the extinction of nearly all marine plankton, 15 percent of marine invertebrate families, the large dinosaurs, and many other species. (Don Davis, NASA)

concentrations on the order of 500 parts per billion. The iridium found in surface sediments typically originates from space and is most often deposited as a light but steady rain of small cosmic particles. Where elevated iridium concentrations are present, extraterrestrial impact may have occurred. (Some scientists favor the alternative theory that iridium has reached the Earth's surface through volcanic activity.) High concentrations of iridium have been found at the Cretaceous-Tertiary boundary in rock and sediment samples from around the world. While iridium anomalies have also been noted in association with other boundaries marking major extinction events, the correlations are less striking and more problematic. To determine iridium concentrations in a sample, chemists use neutron activation analysis, whereby material to be analyzed is exposed to neutrons from a nuclear reactor. Once the iridium has been made radioactive, a gamma-ray detector system determines the element's concentration.

Researchers have identified other signs of possible extraterrestrial impact, all within Cretaceous-Tertiary boundary sediments. Grains of quartz and other minerals have undergone shock metamorphism, a change in crystal structure and density that results from the passage of high-pressure shock waves. Also present are tektites, glassy silicate-rich balls of varying composition, which are believed to form when rock and soil vaporized by an extraterrestrial impact recondense and fall back to Earth. Soot in boundary-layer samples from around the world suggests extensive burning. In addition, geophysical methods have revealed a huge, half-submerged, 65-million-year-old crater, Chicxulub, off Mexico's Yucatán Peninsula. An impact structure almost 180 kilometers in diameter, Chicxulub Crater is buried beneath roughly 2 kilometers of sediment and was detected during oil exploration by means of magnetism and gravity surveys.

To determine whether extraterrestrial impacts are periodic in nature, paleontologists have compiled vast amounts of data on the fossil record of life, notably data on when organisms appeared and disappeared. Researchers noted a possible periodicity in extinction events after creating graphs and conducting statistical analyses using computerized data indexes. Some have also performed similar computer analyses using a list of known impact craters and their ages; periodicities of 28 million to 31 million years have been reported. It should be noted, however, that there is considerable uncertainty in the dates assigned to many of the craters.

Scientific evidence suggesting impact-related extinction and periodic extinction episodes led scientists to develop the Nemesis and Planet X theories. Both theories assume that the extraterrestrial agents in question are comets. Asteroids, rocky astronomical bodies smaller than planets, can also strike the Earth and wreak havoc. However, their travels are considered too erratic to account for the periodicity discussed here. To validate either the Nemesis or Planet X theory, scientists must find an astronomical body whose observed characteristics correlate with predictions.

The Planet X theory is generally considered to be the weaker of these two proposed models. In 1985, when the theory was first published, the existence of a massive tenth planet provided a viable explanation for minor irregularities in the movements of the outer planets. (Pluto's mass proved insufficient to account for the anomalies.) Researchers suggested that a tenth planet had escaped detection because efforts to find it had concentrated on the plane of the solar system, while Planet X's orbit was inclined in relation to the plane. By the mid-1990's, these arguments were less convincing. Although astronomers have an idea where to look, they have yet to observe Planet X. In fact, the lack of orbital irregularity in the outer planets makes it unlikely that a planet massive enough to perturb comets lies in the area predicted by the Planet X theory.

The search for a star matching Nemesis's description continues. Because more than half the known stars are believed to have a companion, the existence of a companion to the Sun would not be unusual. If this companion were a red dwarf star, too small and too dim to be observed readily despite its relative proximity to the Sun, it could escape detection. Moreover, because astronomers have yet to determine the distances to many red dwarfs, it is possible that Nemesis has been observed but not recognized. At present, Nemesis would be halfway through its orbital cycle, near its greatest distance from the Sun (and roughly 13 million years away from its next impact-triggering passage through the Oort Cloud). Researchers are scanning the skies using a computer-controlled reflecting telescope to automatically record and compare images of red dwarf stars. This computerized system observes each candidate over time and uses parallax shift to assess how far away it is. Parallax is the apparent displacement of an object against a background of more distant objects when it is viewed from a different location; the movement of the Earth in its orbit provides the necessary change in viewpoint. The closer an object is to the viewer, the greater the parallax; because Nemesis should be the closest star to Earth other than the Sun, it should exhibit greater parallax than others. So far researchers surveying the northern hemisphere have eliminated more than half of the 3,100 red dwarf stars under scrutiny. The search will continue until astronomers have eliminated all candidates or until Nemesis is found.

Context

The Nemesis and Planet X theories both arose in response to other theories dealing with the nature of extinction on Earth. The idea that collision with an extraterrestrial object could trigger mass-extinction events had been considered before 1980; however, that year saw the first publication of compelling evidence supporting an impact theory. A multidisciplinary research team reported in the journal *Science* that high concentrations of iridium detected in rocks at the Cretaceous-Tertiary boundary were possible evidence that an extraterrestrial agent had ended the reign of the dinosaurs. Geologist Walter Alvarez, physicist Luis Alvarez, and nuclear chemists Frank Asaro and Helen V. Michel proposed that an asteroid or comet roughly 10 kilometers in diameter struck the Earth 65 million years ago, severely affecting the environment and depositing iridium. Their post-impact scenario, according to which the vast quantities of dust and gases flung into the air affected global photosynthesis and temperatures, caught the attention of researchers contemplating the effects of nuclear warfare. They applied the Alvarez team's projections in developing a scenario similar to the "nuclear winter" model advanced by Cornell astronomer Carl Sagan in the early 1980's.

In 1984, paleontologists David Raup and J. John Sepkoski, Jr. published their findings on the periodicity of extinctions. Using Sepkoski's extensive compendium of fossil-record data, the research team was able to distinguish some minor extinction episodes from background extinction levels. They found that over the past 225 million years significant extinction events had apparently occurred approximately every 26 million years.

The scientific community responded quickly with ideas that united and relied upon both the impact and periodic-extinction theories. In April, 1984, the journal *Nature* featured several articles exploring the astrophysical aspects of Raup and Sepkoski's 26-million-year periodicity. Among them was a paper by Walter Alvarez and physicist Richard Muller, who found a 28.4-million-year periodicity in large, well-dated impact craters. In a separate article that gave Nemesis its name, Muller, along with astronomers Marc Davis and Piet Hut, proposed the existence of a small, dim companion to the Sun, a "death

Did an Asteroid Kill the Dinosaurs?

In 1977, Luis W. Alvarez and his son Walter began an investigation of a single-centimeter layer of clay that was sandwiched between two limestone strata containing large deposits of Cretaceous-Tertiary fossils. Such fossils were significantly absent elsewhere in the sample. The clay deposit dated from the boundary between the Cretaceous and Tertiary periods, referred to as the "K-T boundary," roughly 65 million years ago, when the dinosaurs disappeared and modern flora, apes, and large mammals appeared., Alvarez and his son used a trace of iridium in the composition of the clay sample to determine how long it had taken for the clay to be deposited and so to calculate the time that had elapsed during the Cretaceous-Tertiary transition. Iridium is basically an extraterrestrial substance. All the iridium in Earth's crust is only one ten-thousandth of the iridium abundant in meteorites. Alvarez selected iridium because it was the best material to use in determining the amount of debris that fell on Earth during this crucial period. Iridium is deposited uniformly around Earth, and Alvarez wanted to account for these uniform deposits. He began with the theories of Sir George Stokes, who formulated the viscosity law, a calculation of the rate at which small particles fall in the air. Stokes had based this law on his observations of the fallout of ash from the huge eruption of the Krakatoa volcano near Java in the 1880's. After discounting many possible hypotheses, such as a gigantic volcanic eruption, a supernova, or Earth's passing through a cosmic cloud of molecular hydrogen, Alvarez developed the hypothesis that an asteroid had collided with Earth.

According to his calculations, the asteroid had to be 10 kilometers in diameter. Its impact would have been catastrophic, far exceeding the worst nuclear scenario yet proposed. As Alvarez said,

> The worst nuclear scenario yet proposed considers all fifty thousand nuclear warheads in U.S. and Russian hands going off more or less at once. That would be a disaster four orders of magnitude less violent than the K-T asteroid impact.

Alvarez knew that the margin for error in discoveries was exponential, because of the possibility of mistakes in the data. As more data were collected from other sources, however, the argument only became stronger. Although the asteroid hypothesis has not been fully accepted by the scientific community, a number of predictions based on the theory have been verifie experimentally and by computer simulation.

star," that would periodically trigger comet showers on Earth.

Astronomers Daniel Whitmire and Albert Jackson independently reached the conclusion that a distant solar companion drives the extinctions. In January, 1985, Whitmire and astronomer John Matese published a paper in *Nature* that presented Planet X as the driving force for periodic extinction events. Since then, scientists have continued to investigate the possible connection between extraterrestrial impact and extinction, studying the distribution and significance of elevated iridium concentrations, exploring the idea of periodic extinction, and searching for the mechanisms that may drive mass-extinction events.

Nemesis, Planet X, and related theories have generated considerable controversy and interest within the scientific community. Geologists and paleontologists learn that "the present is the key to the past" and that the natural, often gradual forces observable today can explain past geologic events. Hence, to many of these scientists the idea of sudden, violent, global catastrophe, the likes of which humankind has never experienced, seems more like science fiction than sound science. These theories also challenge the notion that evolution's primary driving force is an internal force—competition among species. If cometary impact does indeed occur every 26 million years, external forces such as environmental change may play a more significant role in evolution than scientists previously assumed.

The connection of an asteroid impact (at Chicxulub Crater) to the mass extinction at the Cretaceous-Tertiary boundary is well accepted within the scientific community. However, in the early twenty-first century, some researchers proposed an asteroid impact to explain the even greater mass extinction, often referred to as the Great Dying, that occurred 248 million years ago at the Permian-Triassic boundary. A crater much larger than Chicxulub, buried under 1.5 kilometers of ice in Antarctica, is dated to about 248 million years. That crater also is positioned such that the Siberian Traps are located at the antipode of this crater. This hypothesis suggests that seismic energy from the enormous impact in Antarctica underwent antipodal focusing through the Earth's core to devastate the area now known as the Siberian Traps. That area also suffered tremendous volcanic activity about 248 million years ago. It must be said that this scenario was quite controversial in 2008, but then so was the original Alvarez concept nearly three decades earlier. More research is required before the Permian-Triassic mass extinction in which over 95 percent of all life on Earth died out might be explained in this fashion.

Nemesis and Planet X theories lost favor with time. A number of researchers pointed out that a red dwarf far from the Sun in a rather unstable orbit would likely escape into interstellar space after only a few orbits. Continued observations failed to find a planet beyond Neptune large enough to account for the disruption of small bodies that then enter the inner solar system.

It must also be pointed out that in 2006 the International Astronomical Union approved a new classification system for bodies in the solar system. As a result, Pluto was demoted to the status of a dwarf planet. This left the solar system officially with eight full-fledged planets. Thus the "X" in the Planet X theory could still stand for unknown but would be incorrect if associated with a yet-to-be-discovered tenth planet beyond Pluto.

Karen N. Kähler

Further Reading

Close, Frank. *Apocalypse When? Cosmic Catastrophe and the Fate of the Universe*. New York: William Morrow, 1988. Chapter 4 discusses Nemesis and Planet X. Chapter 5 deals with the Cretaceous-Tertiary extinction and the Alvarez team's research. Includes a glossary. Written for a nontechnical audience.

Dauber, Philip M., and Richard A. Muller. *The Three Big Bangs*. Reading, Mass.: Addison-Wesley, 1996. Section I, which focuses on catastrophic impacts, devotes a chapter to Nemesis and related theories concerning mass extinction. An accessible, well-written book suitable for the general reader.

Goldsmith, Donald. *Nemesis*. New York: Walker, 1985. An accessible book exploring extraterrestrial impacts, the possibility of mass-extinction cycles, and the scientific controversy surrounding the Nemesis and Planet X theories. For the general reader.

Gould, Stephen Jay. *The Flamingo's Smile*. New York: W. W. Norton, 1985. Includes well-written essays addressing Nemesis and related theories (chapter 30), periodic extinction (chapters 15 and 28), and extraterrestrial impacts (chapter 29). For the general reader.

Gribbin, John, and Mary Gribbin. *Fire on Earth*. New York: St. Martin's Press, 1996. Chapter 8, which discusses Nemesis, Planet X, and the basis for periodic-impact theory, presents detailed arguments for and against the Nemesis theory. A thorough and accessible work intended for the general reader.

McBride, Neil, and Iain Gilmour, eds. *An Introduction to the Solar System*. Cambridge, England: Cambridge University Press, 2004. A complete description of solar system astronomy suitable for an introductory college course but also accessible to nonspecialists. Filled with supplemental learning aids and solved student exercises. A Web site is available for educator support.

Muller, Richard. *Nemesis*. New York: Weidenfeld & Nicolson, 1988. Written by a member of the research team that gave Nemesis its name, this book is an engaging firsthand account of the development of the Nemesis theory and its impact on the scientific community. For the general reader.

Raup, David. *The Nemesis Affair*. New York: W. W. Norton, 1986. An accessible account of the genesis of the Nemesis theory and its reception by the scientific community, by one of the chief proponents of the periodic extinction theory. Suitable for the general reader.

Reid, Neil, and Suzanne Hawley. *New Light on Dark Stars: Red Dwarfs, Low-Mass Stars, Brown Stars*. 2d ed. New York: Springer Praxis, 2005. A technical description of those stars that are not very luminous. Discusses discoveries of brown dwarfs and extrasolar planets.

PLANETARY ORBITS: COUPLINGS AND RESONANCES

Category: Planets and Planetology

Orbital or rotational (spin) motions of objects are said to be "coupled" or "in resonance" when the relationships between the periods of such motions can be expressed as ratios of small integers such as 1:1, 1:2, 1:3, 2:3, or 3:4. This usually occurs as the result of the gravitational interaction of the objects. Many examples of couplings and resonances occur in the orbital and rotational motions of solar system objects.

Overview

The orbital or rotational (spin) periods of many solar-system objects have been found to be related by ratios of small integers, such as 1:1, 1:2, 1:3, 2:3, or 3:4. Such relationships in the motions are called couplings or resonances. Usually they have developed over time as the result of gravitational interactions between the objects.

Because of their ubiquity, couplings and resonances are thought to have played a major role in shaping the structure of the solar system.

The many couplings and resonances in our solar system can be categorized into two main types. One type, called a spin-orbit resonance, is manifested by a simple ratio between an object's period of spinning (rotating) on its axis and its period of orbiting (revolving) around a more massive body. For example, the time it takes the Moon to rotate on its axis exactly equals the time it takes to revolve around the Earth, a ratio of 1:1. As a result, the Moon always keeps the same side facing the Earth. (Because the Moon's orbit around the Earth is not precisely circular, its orbital speed varies slightly. Consequently it appears to us on Earth as if the Moon rocks a bit from side to side—a motion called libration—so we end up seeing slightly more than half the Moon's surface during one of its orbits.) Another example is that Mercury's period of rotating on its axis is exactly two-thirds of its period of revolving around the Sun, a ratio of 2:3. Thus, Mercury spins three times on its axis during two orbits around the Sun.

The other type of resonance, called an orbital resonance, involves small-integer ratios between the orbital periods of two or more small-mass bodies orbiting around a much more massive object. Such relationships reinforce the gravitational interactions between the resonant objects. Suppose one of the orbiting small-mass bodies is much less massive than the other small-mass body. (Examples include an asteroid and Jupiter as both orbit the Sun, and a ring particle and a satellite as both orbit Saturn.) The repeated gravitational tugs of the more massive orbiting body (Jupiter or the satellite) on the less massive orbiting body (the asteroid or the ring particle) will tend to pull the less massive body away from its resonant orbit, while the more massive orbiting body is little affected. This clearing of resonant orbits produces gaps in belt and ring systems. Such gaps, or divisions, are observed in the asteroid belt (the Kirkwood gaps), located mainly at the resonances with Jupiter, and in Saturn's ring system as a result of resonances with some of Saturn's satellites.

When two or more objects have exactly the same orbital period around a more massive object, they are called coorbital, a special case of orbital resonance. In 1772, Joseph-Louis Lagrange mathematically discovered five points at which coorbital bodies could exist in equilibrium. These points are called the Lagrangian points and are labeled L1 through L5. Three of the points (L1 through

L3) are unstable, in that an object displaced slightly from the point will drift farther away. However, the L4 and L5 points are stable, in that an object displaced slightly from the point will remain nearby and oscillate around the equilibrium position. These two stable points are located 60 degrees ahead (L4) and 60 degrees behind (L5) the second-largest body along its orbit around the largest body. The Trojan asteroids (so called because they have been named after the heroes, both Trojan and Greek, of the Trojan War) oscillate around the L4 and L5 points 60 degrees ahead of and behind Jupiter along its orbit around the Sun.

Applications

Several types of resonance manifest themselves in the solar system. In all cases, gravity provides the coupling force, although the way gravity is applied to cause the resonance varies. In the cases of spin-orbit resonance, gravitationally produced tides cause the resonances. In cases of orbital resonance, the gravitational forces from two bodies combine to produce either resonant gaps or stable, coorbital points where small particles accumulate. Examples of all these types of resonance can be found within the solar system.

The most familiar example of spin-orbit resonance is the motion of the Earth's only natural satellite, the Moon. Tidal stresses on the Moon from Earth have locked the Moon in its spin-orbit resonance so that only one side faces Earth. Earth dwellers are inclined to think that the Moon does not spin or rotate, but if viewed from far out in space, the Moon would be seen to spin once for every orbit it makes of the Earth. As an illustration, the Moon has phases because sunlight reaches all points of the Moon. This indicates that, as viewed from the Sun, the Moon spins once a month, which is exactly the same time that it takes to orbit the Earth, and is therefore in a 1:1, or synchronous spin-orbit, resonance with Earth. The Moon is often said to be "tidally locked" to the Earth because of this resonance and its cause.

The Moon is not the only secondary satellite in the solar system to exhibit synchronous rotation. Tidal locking appears to be the rule for all satellites close to a planet. In fact, Phobos and Deimos are in synchronous rotation around Mars. The four Galilean moons of Jupiter, which are some of the largest satellites in the solar system, also exhibit a 1:1 spin-orbit resonance. Of the many other satellites of Jupiter, two others that had their rotational periods measured; one of those, the closest one to Jupiter, is synchronous. A similar situation exists for the satellites

of Saturn, where eight are known to have synchronous rotations; eight other sizable satellites have not been measured. At least two Saturnian satellites are nonsynchronous. The largest satellites of Uranus and Neptune are also in synchronous rotation, and the smaller ones have not yet been measured. Pluto's companion Charon not only is in synchronous rotation around Pluto but also is large and close enough to have caused Pluto's rotation to be synchronous with Charon's orbit. Thus, all planets with satellites have examples of synchronously rotating moons.

Mercury, although lacking any moon, also exhibits spin-orbit resonance. Mercury spins three times for every two orbits it makes of the Sun. This 3:2 spin-orbit resonance is related to the unusually elongated, elliptical orbit of Mercury. As a result of its resonance condition, whenever Mercury is at perihelion, the same point on the planet is either facing directly toward or away from the Sun. The Mariner 10 spacecraft identified a huge feature, the Caloris Basin, at this point on the surface and some strange surface features, dubbed Weird Terrain, on the planet's opposite side. These discoveries suggest that a huge, ancient impact that nearly tore the planet apart made one side of the planet heavier than the other and probably elongated and tilted the orbit. Tidal effects over the years have slowed the rotation of the planet so that whenever Mercury is at perihelion, its heavy side points either toward or away from the Sun, as a tidal bulge would. The fact that the spin is not synchronous with the orbit, as it is for Earth's moon, is most likely the result of Mercury's large mass and elongated orbit, which brings it considerably closer to the Sun at perihelion than at aphelion.

Evidence of orbital resonance was discovered in the asteroid belt between Mars and Jupiter in 1866 by Daniel Kirkwood. As he studied the orbits of the asteroids, Kirkwood discovered gaps in an otherwise congested region of space. Since Kirkwood's original discovery of gaps at 2:1, 3:1, and 4:1 resonances with Jupiter, at least five other gaps have been identified. It is apparent that the strong and repeated pull of Jupiter destabilized orbits of asteroids with these periods and opened up the Kirkwood gaps.

Divisions in Saturn's rings have a cause similar to the Kirkwood gaps. Gian Domenico Cassini first observed the largest gap, the Cassini division, in 1675; in 1867, Kirkwood discovered that the Cassini division has a 2:1 orbital resonance with the satellite Mimas. Kirkwood also showed that the Cassini division was in a 3:1 resonance with the satellite Enceladus, a 4:1 resonance with the moon Tethys, and a 6:1 resonance with the satellite Dione, although the Mimas resonance is probably more significant because that coupling is stronger and more frequent. In addition to the Cassini division, there are gaps in the A ring at resonances with the satellites Janus (S10) and Epimetheus (S11). Moreover, the edges of the A and B rings, which are very well defined, occur at resonance locations.

There are still some mysteries to be found in the asteroid belt and Saturn's rings. Surprisingly, at the Jupiter 3:2 resonance location in the asteroid belt, there is accumulated material instead of an empty gap. In the Cassini division, there are ringlets, which may be spiral density waves excited by resonances with the satellite Iapetus. Other details of the structure and shape of the divisions are still not well understood and may be aspects of density waves and chaotic behavior. Nevertheless, these are details, and the main features must be caused by the simple resonances.

Jupiter and Saturn also have several examples of coorbital satellites. In Jupiter's orbit around the Sun, there are clumps of asteroids one-sixth of an orbit ahead and behind Jupiter at the Lagrangian points. These coorbital asteroids are called Trojan asteroids. In the Saturnian system, the satellite Tethys has two Lagrangian coorbital satellites, Telesto (following Tethys) and Calypso (leading Tethys). In addition, the satellite Dione has a coorbital satellite named Helene at the leading Lagrangian point. Other examples are expected to exist, but they have not yet been observed.

Two of Saturn's satellites, Janus and Epimetheus, are also coorbital, but in a different way from that of the Lagrangian coorbitals. Janus and Epimetheus have orbits that are so close together that their gravitational attraction for each other is sufficient for them to interchange orbits without colliding. The difference between this case and that of the Lagrangian coorbital satellites results from the fact that Janus and Epimetheus are nearly the same size and that one satellite does not always lead the other.

Context

The result of the Moon's spin-orbit resonance has been known ever since humans became aware of the world around them, but the mechanism for understanding why such a resonance would occur was not discovered until Sir Isaac Newton formulated his laws. In fact, tides were explained in his *Philosophiae Naturalis Principia Mathematica* (1687; *Newton's Principia: The Mathematical Principals of Natural Philosophy*, 1846), in which he first used his laws publicly to explain Johannes Kepler's laws of planetary motion. It was thought that most satellites would have similar resonances with their planets, but this extrapolation needed to be confirmed. Planetary probes have visited all planets except Pluto, and the New Horizons spacecraft is on its way to a flyby of the Pluto-Charon system in 2015. With a few notable exceptions, many satellites have exhibited spin-orbit resonance. Unfortunately, all the satellites were not able to be studied thoroughly to determine their spin rates; as a result, a complete understanding of which moons are tidally locked to their planets, and how, will not be within reach until additional probes or improved technology becomes available. The Galileo probe improved our understanding of Jupiter's satellites, and Cassini added to what is known about Saturn's.

Interestingly, Mercury generally was thought to be tidally locked to the Sun ever since Giovanni Schiaparelli made crude maps of the surface in the 1880's. This opinion seemed to be confirmed by later Earth-based observations that were carried out prior to the early 1960's. It was not until Doppler radar techniques were applied to Mercury in 1965 that the 3:2 spin-orbit resonance was discovered. In this study, radar signals were sent from the 300-meter Arecibo radio telescope in Puerto Rico and bounced off Mercury. The change in the signal's frequency (the Doppler effect) proved that Mercury rotates once in 58.65 Earth days instead of the 88 days that it takes to orbit the Sun. It was not until the three Mariner 10 flybys of Mercury in 1974 and early 1975 that the Caloris Basin and the Weird Terrain were discovered and the orbital resonance was confirmed.

Discovery of the Cassini division in 1675 also predates Newton's laws, which are essential to explain the division. It was not until Kirkwood's discovery of resonance conditions in 1866 and 1867 that a reasonable explanation for formation of this division was offered. Kirkwood's model for the Cassini division and gaps in the asteroid belt was thought to be adequate until the Voyager data became available in 1981. Images of the Cassini division from the two Voyager spacecraft revealed a number of unexpected details. These details require a more sophisticated application of Newtonian mechanics and provide a testing ground for density wave theories and theories of chaotic behavior. These theories could help explain the structure of galaxies.

In contrast to previous cases, Lagrange predicted in 1772 the location of stable, coorbital companions to Jupiter and other planets. In 1906, Jupiter's coorbital companions were found. The search continues with nearly two hundred such asteroids discovered to date and perhaps ten times that number orbiting at Jupiter's Lagrangian points. The coorbital satellites in Saturn's system were undiscovered until the Voyager flybys in 1980 and 1981. In fact, Lagrangian coorbital satellites may exist for Earth, Mars, and other large bodies.

Larry M. Browning

Further Reading

Elliot, James, and Richard Kerr. *Rings: Discoveries from Galileo to Voyager*. Cambridge, Mass.: MIT Press, 1984. A detailed discussion of the structure and formation of planetary rings. Discusses the connection between the Cassini division, spiral density waves, and galaxies.

Fowles, Grant R., and George L. Cassiday. *Analytic Mechanics*. 7th ed. New York: Brooks/Cole, 2004. A college textbook for a second course in Newtonian mechanics. Particularly strong on theory and applications of orbital motion. Requires knowledge of advanced calculus.

Karttunen, H. P., et al., eds. *Fundamental Astronomy*. 5th ed. New York: Springer, 2007. A well-used university textbook in introductory astronomy. Contains some calculus-based treatments for those who find the standard treatise for typical ASTRO 101 classes too low level. Suitable for an audience with varied science and mathematical backgrounds. Covers all topics from solar-system objects to cosmology.

Miner, Ellis D., Randii R. Wessen, and Jeffrey N. Cuzzi. *Planetary Ring Systems*. New York: Springer Praxis, 2006. Perhaps the most comprehensive text on planetary ring systems that is also accessible to the general reader. Provides interpretation of Pioneer, Voyager, Galileo, and Cassini data and observations. Extensive notes, tables, figures, and references. Accessible to a nonexpert, scientifically inclined audience.

Shu, Frank H. *The Physical Universe: An Introduction to Astronomy*. Mill Valley, Calif.: University Science Books, 1982. Although somewhat dated and more mathematical than the other references, this book contains an extensive and enlightening discussion of resonances in the solar system. Also noteworthy is the discussion of spiral density waves and their application to the Cassini division by one of the pioneers of that theory.

Time-Life Books. *Comets, Asteroids, and Meteorites*. Alexandria, Va.: Author, 1990. This volume discusses the Trojan asteroids orbiting with Jupiter and the theories about the formation of the Kirkwood gaps. Very readable and well illustrated.

_____. *The Far Planets*. Alexandria, Va.: Author, 1988. Notable for its pictures and informative illustrations of Saturn's ring system. Published before the August, 1989, flyby of Neptune by Voyager 2.

_____. *The Near Planets*. Alexandria, Va.: Author, 1989. This volume provides a very readable summary of the exploration of Mercury and its spin-orbit resonance. A good source of scientifically accurate illustrations.

Wagner, Jeffrey K. *Introduction to the Solar System*. Philadelphia: Saunders College Publishing, 1991. A well-written and up-to-date discussion of all aspects of the solar system, especially spin-orbit coupling.

PLANETARY SATELLITES

Categories: Natural Planetary Satellites; Planets and Planetology

All but two of the planets in the solar system have one or more smaller bodies in orbit around them. These satellites, popularly called "moons," are important for what they tell scientists about the origins of the planets and the evolution of the solar system.

Overview

Of the planets in the solar system, all but two, Mercury and Venus, have at least one satellite. There are more than one hundred known satellites in the solar system, ranging in size from about 15 kilometers to more than 5,200 kilometers in diameter; the number continues to grow as spacecraft and telescopic observations discover more and more relatively tiny satellites about the gas giants, particularly Jupiter and Saturn. Although the largest satellites of the solar system's eight planets have been identified, a table of known satellites would become out of date within a matter of months; more are being discovered all the time. Satellites are important scientifically for the clues they provide to the origin of their parent planets and the solar system in general.

Most models of planetary formation start with some variation on the "nebular hypothesis" devised by German philosopher Immanuel Kant in 1755 and later modified in 1796 by French mathematician Pierre-Simon Laplace. This hypothesis in modern form suggests that the Sun, planets, satellites, and smaller debris in the solar system started as a rotating flattened cloud of gas and dust. This rotating cloud eventually became unstable and broke into rings. Material concentrated in the bulging center of the cloud became compressed, eventually giving birth to the Sun. The gaseous rings cooled eventually, allowing precipitation of solid crystals that aggregated to form rocky materials. These aggregates, in turn, accreted over geological time into larger and larger bodies. The largest of these bodies became the major planets, with smaller bodies generally falling into the planets, adding to their mass. Under certain circumstances, small bodies were captured as satellites or remained freely orbiting the Sun as dwarf planets, asteroids, and comets.

Like the major planets, satellites vary in composition depending on their mean distance from the Sun. Satellite composition can be expressed in terms of two major components: ice and rocky silicate material. Satellites in the inner solar system (Earth's moon, for example) are composed almost exclusively of rocky material, whereas those orbiting Jupiter and beyond are mixtures of ice and rocky material. Thus, satellite density decreases in the outer solar system. The distribution of ice and rock among satellites parallels the preponderance of rocky planets in the inner solar system (Mercury through Mars) and gas-rich planets in the outer solar system (Jupiter through Neptune). Pluto is composed almost completely of ice and has been reclassified as either a dwarf planet or plutoid, the first member of a class of objects from the Kuiper Belt referred to as plutinos. From this distribution, it can be noted that "volatile" substances (those with low melting and boiling points) are concentrated in the outer solar system. "Refractory" substances, those with high melting and boiling points, occur in greater abundance within the inner solar system. Satellites in the solar system reveal a considerable diversity in terms of surface composition and physical characteristics.

Earth's moon, the nearest satellite to the Sun, is a refractory-rich rocky triaxial ellipsoid approximately 3,500 kilometers in average diameter, more than one-fourth the size of Earth. The Moon is the best-known of all other extraterrestrial satellites because of the many spacecraft that have examined it from orbit and landed on its surface. These robotic missions began with the 1959 mission by the Soviet Union to photograph the lunar "far side," followed by the National Aeronautics and Space Administration's (NASA's) successful Ranger, Surveyor,

and Lunar Orbiter projects in the 1960's. The most famous lunar project of course was NASA's Apollo program, which landed humans on the Moon. From 1969 to 1972, six Apollo missions landed twelve astronauts on the Moon, who returned numerous rocks and "soil" samples. The Apollo astronauts also set up experimental equipment for measuring "moonquakes" and other phenomena. Information from all the lunar missions, whether crewed or robotic, indicate that the Moon is a complex, small satellite with a metallic core, an iron-rich silicate mantle, and a silicon plus aluminum-rich silicate crust.

Nearly two decades transpired before another spacecraft was sent to the Moon after the Russian Luna 24 mission returned a small amount of rock and soil samples to Earth from the Moon's Sea of Crises in 1976. By this point, NASA had moved on to attempting to land robotic spacecraft, the Vikings, on Mars. In early 1998 NASA sent the Lunar Prospector into lunar orbit. Outfitted with five different instruments, Lunar Prospector's objectives were to globally map the lunar resources, determine the Moon's complex gravity field and its minor magnetic field, and to look for any outgassing from the surface. The surprising result of Lunar Prospector investigations was the strong indication that between 10 to 100 million metric tons of water ice could be located in permanently shadowed portions of craters near the lunar poles.

The presence of water on the Moon as a resource that could be utilized would greatly enhance the viability of crewed lunar research outposts, especially if those outposts were constructed at or very near the Moon's poles. This finding was not direct evidence. The spacecraft's neutron spectrometer picked up the presence of protons; the most reasonable extrapolation of the data was that those protons were bonded together in water ice. The potential finding of water on the Moon was not a total surprise. Two years earlier, a Ballistic Missile Defense Organization probe called Clementine used the Moon as a testbed for new sensor technologies, among other things. Some of the sensor data hinted at the possibility of water near the poles. Lunar Prospector took the search for water to a new level, and the real surprise was the total amount of water that Lunar Prospector seemed to find.

The Moon has no significant atmosphere, a trait it shares with most other small bodies in the solar system (Saturn's satellite Titan and Neptune's satellite Triton are exceptions). Also, the Moon's crust is far less complex than Earth's. It consists of densely cratered highland areas composed of a feldspar-rich (calcium-aluminum silicate) rock called anorthosite (light-colored areas), with dark

Data on Major Satellites in the Solar System

Planet	Satellite	Diameter (km) or Dimensions	Mass (kg)	Density (g/cm3)
Earth	Moon	3,476	7.35×10^{22}	3.34
Mars	Phobos	$27 \times 22 \times 19$	9.60×10^{15}	2
	Deimos	$15 \times 12 \times 11$	1.90×10^{15}	2
Jupiter	Io	3,630	8.92×10^{22}	3.53
	Europa	3,138	4.87×10^{22}	3.03
	Ganymede	5,262	1.49×10^{23}	1.93
	Callisto	4,800	1.08×10^{23}	1.70
Saturn	Mimas	392	4.50×10^{19}	1.43
	Enceladus	500	7.40×10^{19}	1.13
	Tethys	1,060	7.40×10^{20}	1.19
	Dione	1,120	1.05×10^{21}	1.43
	Rhea	1,530	2.50×10^{21}	1.33
	Titan	5,150	1.35×10^{21}	1.89
	Iapetus	1,460	1.88×10^{21}	1.15
	Phoebe	220	?	?
Uranus	Miranda	480	7.50×10^{19}	1.26
	Ariel	1,330	1.40×10^{21}	1.14
	Umbriel	1,110	1.30×10^{21}	1.82
	Titania	1,600	3.50×10^{21}	1.60
	Oberon	1,630	2.90×10^{21}	1.28
Neptune	Triton	3,800	1.30×10^{23}	?
	Nereid	300	2.10×10^{19}	?
	1989 N1	400	?	?
Pluto	Charon	1,150?	?	2 ?

Note: Jupiter, Saturn, Uranus, and Neptune also have other, minor satellites. Venus and Mercury have no known satellites. Pluto, designated a dwarf planet by the International Astronomical Union in 2006, is considered by some to form a doubleplanet system with Charon.

basalt lava flows (iron-rich silicate rock) filling huge impact craters that the Italian astronomer Galileo named *maria* (Latin for "seas"). The Moon contains no real granite rocks such as those that compose much of Earth's continents.

The Moon always shows the same face to Earth as it orbits, because the Moon's rotational rate is equal to its orbital rate. This situation is an example of "synchronous rotation," in which the rotation rate of a body has some precise mathematical relationship to orbital period (time

required to complete one orbit). The Moon's 1:1 ratio of orbital-to-rotational period results from the Earth-side of the Moon bulging out because of gravitational tidal forces between Earth and the Moon. Eventually, this bulge (the side facing Earth) comes to lie along the Earth-Moon line, its most stable configuration. Other bodies in the solar system show similar relationships. For example, the satellites of Mars, Phobos and Deimos, rotate so that the same side always faces Mars. Many satellites of Jupiter and Saturn also show this relationship. Pluto and its satellite

Charon both revolve and rotate at the same rate. This means that an observer on Pluto would always see Charon in precisely the same place in the sky all day and night.

Traveling out from the Sun, the next planet is Mars, with its tiny satellites Deimos and Phobos. These oddly shaped, rocky bodies (neither is spherical) are most likely escapees from the nearby asteroid belt, a zone between Mars and Jupiter that contains thousands of small planetoids (up to about 1,000 kilometers in diameter), rock fragments, and dust. Photographed up close by the Viking Orbiter 1 in 1977, Phobos and Deimos appear to be composed of the same materials that occur in certain meteorites.

At Jupiter, a miniature solar system is found. Jupiter and most of satellites were photographed extensively by Voyagers 1 and 2 in 1979 and earlier by Pioneer 10. Jupiter's four largest satellites are called the "Galilean satellites" because they were discovered by Galileo in 1610. Their densities and rock-to-ice ratios decrease with increased distance from Jupiter, a relationship that is mirrored by the larger solar system. In addition, geological activity decreases from the closest Galilean moon, volcanically active Io, to the highly cratered outermost satellite, Callisto. The high crater density on Callisto's surface shows that its surface is very old and has not become "resurfaced" by high-energy processes such as volcanic activity or erosion. This same reasoning is used elsewhere in the solar system to deduce relative ages of satellite and planetary surfaces. Earth's moon, Mercury, and many other bodies are heavily cratered by impacting projectiles and, therefore, are considered to have old surfaces. The Galilean satellites Io and Europa show no craters because their surfaces are renewed constantly by molten sulfurous compounds on Io and liquid water that freezes to ice on the surface of Europa. The heat source to produce this volcanic activity results from tidal forces originating from massive Jupiter.

Saturn and most of its sixty satellites were photographed and studied by Pioneer 11, the two Voyager spacecraft, and the Cassini orbiter. Most of its satellites are icy, heavily cratered worlds; one satellite, Titan, is the second largest satellite in the solar system (Jupiter's Ganymede is larger). Titan appears to have a surface composed of complex hydrocarbon compounds and liquid nitrogen, but it is unusual for such a small "planet" to have its extensive atmosphere, which is composed mostly of methane (CH_4). Its presence may be caused by the extremely cold temperatures or by the continual production of gases by some type of its cryogenic volcanic activity.

Uranus, visited only by Voyager 2 in 1986, has at least twenty-four satellites, of which only five were large enough to have been observed from Earth prior to Voyager's visit; many were discovered long after that visit. Of all the satellites photographed by Voyager 2, the most surprising by far is Miranda. Miranda has nearly crater-free dark areas with concentric grooves and chevron-shaped features that are separated by huge fault scarps (cliffs) from heavily cratered areas. Proposals to explain Miranda's bizarre surface features included the breakup and reassembly of the satellite following a catastrophic collision with another body or the upwelling of heated water to form its dark, bulging, grooved terrains.

Neptune was visited in 1989 by Voyager 2 and its two major satellites, Nereid and Triton, were photographed, along with six new ones. One of those, 1989 N1, measures about 400 kilometers in diameter, replacing Nereid as the second largest Neptunian satellite. Nevertheless, Triton is the most important. It shows a frozen surface with virtually no impact craters. Triton's landscape includes huge frozen "lakes" of solidified water mixed with ammonia, along with a so-called cantalouped terrain of intersecting grooves and ridges that may represent fault systems. Frigid Triton has a surface tempe.rature of only 37 kelvins. It has scattered, dark linear streaks on its surface that may represent nitrogen-powered geysers spewing dark organic matter out on the surface. Winds blowing in Triton's thin methane atmosphere align these geyser streaks and blow thin, ice crystal clouds around this miniature planet. Triton is a unique world that has somehow maintained internal heating and recent resurfacing. It is truly a remarkable feat for a body in such a cold place.

Applications

Satellites are studied primarily for what they tell scientists about the origin of the solar system. As intelligent beings who are products of solar system evolution, humans have a natural curiosity about the solar system's and our own origin. The mechanisms of satellite formation and eventual capture by larger planets illustrate many of the dynamic processes that have shaped the solar system since its condensation from the solar nebula perhaps 4.5 billion years ago.

As an example, Earth's moon poses interesting problems regarding attempts to understand the origin of the Earth. Many ideas have been proposed for the origin of the Earth-Moon system. These ideas include the hypothesis that the Moon formed elsewhere in the solar system and was later captured by the Earth, or that the Moon "fissioned" off from the early Earth itself, possibly leaving a major hole how known as the Pacific Ocean. Detailed analyses of lunar samples returned by Apollo astronauts (1969-1972) and the Soviet robotic Luna missions (which concluded in 1976) show that lunar rocks are similar in some important respects to terrestrial rocks, but they also show critical differences. Lunar and earthly rocks have the same ratios of the three oxygen isotopes (nuclei with the same atomic numbers but different mass number—hence different amounts of neutrons but the same number

of protons): oxygen 15, oxygen 17, and oxygen 18. This indicates that their constituents condensed from the same area in space. Thus, the theory that postulates the Moon was captured after forming elsewhere is no longer held in much esteem.

The fission theory contrasts with the idea that the Moon was formed at some area removed from the Earth. Concentrations of other chemicals in lunar samples than oxygen isotopes display great differences from Earth rocks. For example, the Moon is richer in iron and carbon

Moons of the solar system's planets appear here in composite. (NASA)

compared to Earth and contains far more refractory elements (such as titanium and calcium) compared to volatile substances (such as water, sodium, and potassium). These differences indicate that Earth and the Moon could not have formed from exactly the same starting materials as would be expected if they condensed from a common area of the original solar nebula. Therefore, the fission theory is also no longer considered viable.

An idea to resolve the problem of lunar origin, called the "impactor hypothesis," involves a violent collision of the early Earth with a wayward, Mars-sized planet (the "impactor"). This collision would have blasted out material from Earth all the way to the metallic core and would volatilize or pulverize most of the impactor. A mixture of loose impactor and Earth materials then would have orbited the wounded Earth, eventually accreting (accumulating) to form the Moon. Advocates of this hypothesis argue that mixing of impactor and Earth materials to form both the Moon and Earth would explain the similar oxygen isotope ratios. Yet, because the Moon and Earth would be composed of different ratios of proto-Earth and impactor materials, the two bodies' differences in major chemical components such as water and iron are also explained. Fortunately for the future evolution of life, Earth acquired most of the volatiles, including water, which is essential to initiate and sustain life. This theory presently is preferred.

The complexity of planet and satellite formation is well illustrated by bodies in the outermost reaches of the solar system; the Neptunian system and the Pluto-Charon system are particularly intriguing. Large, icy Triton orbits Neptune in a "retrograde" direction, clockwise when viewed from "above" the ecliptic plane of the solar system. Most planets and satellites rotate and revolve in a counterclockwise direction. Voyager 2 images and measurements show that Triton and Nereid are very similar to Pluto-Charon in composition and density. All are methane-rich ice balls that resemble the comets that originated in the outer solar system. Theorists speculate that Triton, Nereid, and perhaps Pluto may be comet-like objects that were captured by Neptune early in the history of the solar system and revolved around Neptune in the normal direction. Triton's retrograde orbit may have resulted from a collision or close encounter with a large passing planetoid, an event that may also have caused Nereid to assume its strange, elongated elliptical orbit. In this model, Pluto was split in two (to make Charon) by the encounter and ejected into the outer solar system. Regardless of whether this hypothesis is correct, the

Neptunian system is known to be unstable, indicating that something has disturbed it since the original formation of the solar system. In about 10 to 100 million years, Triton will spiral close enough to Neptune to be torn to pieces by tidal forces, adding greatly to the mass of Neptune's current thin system of partial arcs.

Although most study of planetary satellites is conducted to learn more about the solar system in general and the possible future of Earth in particular, knowledge of one nearby satellite, Earth's moon, will eventually be used to practical advantage. For example, dark lava flows (basalt) on the Moon contain abundant titanium, a valuable component in high-temperature metal alloys (like metal used in rocket bodies). Rocks in the lunar highland areas are mostly anorthosites, composed primarily of the mineral plagioclase feldspar, which is a potential source of aluminum. Some aluminum is extracted from feldspar deposits on Earth, but the process to separate aluminum from the rest of the mineral requires a high-energy input. On Earth, this process is accomplished commonly by locating the aluminum smelter near a source of hydroelectric power; on the Moon, the energy would have to come from the Sun or a small local nuclear power plant.

Earth's Moon is a potential base for launching spacecraft to other regions of the solar system. This endeavor would require the construction of a lunar base where the Moon's low gravity would greatly facilitate spacecraft launches. Far less energy must be expended on the Moon to escape its gravity field than is required on Earth. Wholesale colonization of the Moon to alleviate population pressures on Earth, however, is not feasible. One aspect of the scientific study of the Moon confirms the distinct lack of concentrated water supplies on the Moon, although some researchers believe that subsurface water in localized "permafrost" deposits will be discovered in the future, such as those found on Mars. Even if they exist, these frozen water deposits would not support large populations of humans.

In 2004, in the aftermath of the *Columbia* accident, President George W. Bush charged NASA with fulfilling his administration's Vision for Space Exploration, which called for a permanent human outpost on the Moon and development of technology to expand the human presence to Mars and beyond. NASA pursued the Constellation program as a versatile crewed transportation system to return astronauts to the Moon, build a base there near the South Pole where the water appears to be located in greatest abundance, and then work toward eventual trips to Mars. A nonbinding deadline for reaching the Moon was set at 2020.

However, the United States was not the only nation with an interest in sending humans back to the Moon. The Chinese had already announced their plans for sending taikonauts (the Chinese word for astronaut) into space and in due course to the Moon. The Europeans sent the Small Missions for Advanced Research in Technology 1 (SMART 1) spacecraft to the Moon to produce the highest-resolution mapping of its surface. A Chinese satellite named Chang'e 1 was sent into orbit, the Japanese sent their Kaguya probe to the Moon, and the Indians launched their Chandrayaan spacecraft on October 22, 2008. For the first time an international fleet of probes was investigating the Moon, and with a government-supported space program, it appeared that Chinese taikonauts might reach the Moon by the second decade of the twenty-first century.

Context

The five planets known to ancient peoples were referred to as the "wandering stars" to distinguish them from the "fixed stars." At least by the time of the ancient Greeks, the Moon was recognized as a satellite of the Earth. Most Greek philosophers, such as Plato and Aristotle, believed that all planets and the Sun revolved around the Earth.

The history of astronomy and of science in general was influenced profoundly by Galileo's discovery of the Jovian satellites. Galileo had constructed a crude telescope and used it to scan the heavens. He discovered the four largest satellites of Jupiter (Io, Europa, Ganymede, and Callisto) now known as the "Galilean satellites" in his honor. In 1610, Galileo published his findings in *Sidereus Nuncius* (starry messenger). Before that, the heliocentric (sun-centered) model of the solar system proposed by Nicolaus Copernicus was seriously questioned and indeed considered heretical by the Catholic Church. Galileo's discovery of a "miniature solar system" consisting of satellites around a planet demonstrated that objects could revolve around something other than the Earth, thus displacing the Earth from the center of the universe.

The next major discovery—the large satellite of Saturn, Titan—was made by the Dutch astronomer Christiaan Huygens in 1656. By the end of the nineteenth century, eight more satellites had been discovered orbiting Saturn, many of them found by the Italian astronomer Gian Domenico Cassini. Huygens also realized that the "ears" on either side of Saturn first described by Galileo were actually rings, a feature attributed to individual particles (tiny satellites) by James Clerk Maxwell in 1857.

Although the distant planet Uranus was discovered in 1781 by English astronomer Sir William Herschel, its five largest satellites required an additional 167 years to detect. The last one, tiny Miranda, was discovered by American astronomer Gerard Peter Kuiper in 1948.

Neptune was discovered in 1846 by German astronomer Johann G. Galle, followed in the same year by the discovery of its large satellite Triton by English astronomer William Lassell. Charon, the largest satellite of Pluto, was discovered in 1978 by James W. Christy of the United States Observatory after he noticed that a photographic image he had taken of Pluto showed a lump on one side. This lump was shown later to move relative to Pluto, confirming the existence of a satellite.

These later discoveries of other satellite systems, along with continuing new discoveries, have proven crucial to the understanding of the origin of the planets, as well as the satellites themselves. Scientific study of satellites shows that the solar system evolved amid violent collisions, gravitational perturbations of orbits, and heating and cooling of surfaces and interiors, in a bewildering variety of combinations. These studies show that, although they share some characteristics, every planet-satellite system formed in some unique way according to conditions prevailing in its particular region of the solar system.

John L. Berkley

Further Reading

Fischer, Daniel. *Mission Jupiter: The Spectacular Journey of the Galileo Spacecraft.* New York: Copernicus Books, 2001. Suitable for a wide range of audiences. Thoroughly explains all aspects of the science and engineering of the Galileo spacecraft. Particularly good are the discussions about the nature of the Galilean satellites.

Greenberg, Richard. *Europa the Ocean Moon: Search for an Alien Biosphere.* New York: Springer, 2005. A complete description of current knowledge of Jupiter's satellite Europa through the post-Galileo spacecraft era. Discusses the astrobiological implications of an ocean underneath Europa's icy crust in a readable text geared for astronomy enthusiasts as well as college students. Well illustrated.

Harland, David M. *Cassini at Saturn: Huygens Results.* New York: Springer, 2007. The text provides a thorough explanation of the entire Cassini program, including the Huygens landing on Saturn's largest satellite, Titan.

Essentially a complete collection of NASA releases from the start of Cassini flight operations through the majority of Cassini's seventy orbits of its primary mission, which concluded a year after this book entered print. Somewhat technical.

_____. *Mission to Saturn: Cassini and the Huygens Probe*. New York: Springer Praxis, 2002. A technical description of the Cassini program, its science goals, and the instruments used to accomplishment those goals. Written before Cassini arrived at Saturn, it provides a historical review of pre-Cassini knowledge of Saturn and its satellites.

Hartmann, William K. *Moons and Planets*. 5th ed. Belmont, Calif.: Thomson Brooks/Cole, 2005. A college-level textbook written in a style accessible to non-science majors. New terms are written in boldface type and defined in the text. Lavishly illustrated with well-executed drawings and black-and-white photographs. Hartmann is a distinguished planetary scientist and an accomplished artist. His paintings of remote planetary scenes have fostered a greater understanding of the variety of features displayed by planetary surfaces. Includes appendices, with extensive planetary vital statistics, a chapter-by-chapter bibliography, and detailed index.

Irwin, Patrick G. J. *Giant Planets of Our Solar System: An Introduction*. 2d ed. New York: Springer, 2006. Suitable as a textbook for upper-level college courses in planetary science. Focuses on Jupiter, Saturn, Uranus, and Neptune and their satellites, rings, and magnetic fields. Filled with figures and photographs.

Lovett, Laura, Joan Harvath, and Jeff Cuzzi. *Saturn: A New View*. New York: Harry N. Abrams, 2006. A coffee-table book replete with about 150 of the best images returned by the Cassini mission to Saturn. Covers the planet, its many satellites, and the complex ring systems.

McBride, Neil, and Iain Gilmour, eds. *An Introduction to the Solar System*. Cambridge, England: Cambridge University Press, 2004. A complete description of solar system astronomy suitable for an introductory college course, but accessible to nonspecialists as well. Filled with supplemental learning aids and solved student exercises. A Web site is available for educator support.

Rothery, David A. *Satellites of the Outer Planets: Worlds in Their Own Right*. New York: Oxford University Press, 1999. Up to date with the Galileo spacecraft results, this heavily illustrated, easy-to-read book offers a comprehensive geological study of the satellites of the gas giants.

Van Pelt, Michel. *Space Invaders: How Robotic Spacecraft Explore the Solar System*. New York: Springer, 2006. A historical account of robotic planetary science missions attempted by all spacefaring nations, written by an European Space Agency cost and systems engineer. The narrative not only explains the science but also provides a behind-the-scenes description of the development of a space exploration mission from concept proposal to flight operation.

PLUTO AND CHARON

Categories: Natural Planetary Satellites; Planets and Planetology; Small Bodies

Pluto became the ninth planet in the solar system when discovered in 1930. Then, in 2006, the International Astronomical Union demoted Pluto to the status of a dwarf planet. Pluto and its satellite Charon constitute a dual-object system located far from the Sun. These bodies are different in size and composition from any of the planets of the solar system, more closely resembling the icy satellites of Neptune.

Overview

Pluto was discovered by American astronomer Clyde Tombaugh in 1930; its satellite Charon was detected in 1978 by James W. Christy of the U.S. Naval Observatory. In the United States, "Charon" is pronounced SHAR-uhn, reminiscent of the discoverer's wife, Charlene; in the rest of the world, the pronunciation KAR-uhn is usually preferred.

Less is known about Pluto than about any of the other planets; it is the only planet that has not been visited by a satellite from the Earth. Earth-based telescopes cannot provide much information about Pluto and Charon, as they are too far away for surface details to appear even in the largest telescopes. Better images of Pluto were obtained using the Hubble Space Telescope, which in May, 2005, detected two new small satellites of Pluto. Hubble images also provided the first indications of features on the surface of Pluto. Almost a dozen "provinces"—portions of Pluto with different albedos—were discovered. Pluto also was observed to have a northern

The Hubble Space Telescope captured this image of Pluto (lower left) and Charon, more than 4 billion kilometers away, in 1994. (Dr. R. Albrecht, ESA/ESO Space Telescope European Coordinating Facility; NASA)

polar cap, several dark spots, and a bright linear feature. The resolution of these Hubble images precluded a precise determination of the nature of these intriguing features. However, scientists speculated that the images were hinting at basins or craters of significant size. Identification of these features would have to wait for the New Horizons spacecraft to fly through the Pluto-Charon system in 2015.

Pluto is the smallest planet in the solar system, smaller even than Earth's moon; it is usually the outermost planet. Pluto takes 247.7 Earth years to orbit the Sun and rotates on its axis once every 6.39 Earth days. The orbit of Pluto is more eccentric than that of any other planet. Pluto's orbital eccentricity is so large that Pluto is sometimes closer to the Sun than Neptune: That was the situation between January 21, 1979, and March 14, 1999, when Pluto's orbit again took it farther from the Sun than Neptune. Pluto will remain beyond the orbit of Neptune until 2226.

Distances between objects in the solar system are usually measured in astronomical units (AU). One astronomical unit is the average distance between the Earth and the Sun, or 150 million kilometers. Pluto can be as close to the Sun as 29.64 AU and as far away as 49.24 AU.

At Pluto's distance, the Sun appears as a starlike point, but a point more than one hundred times brighter than a full Moon. The amount of solar energy received by Pluto varies greatly because of the large variation in distance between the Sun and Pluto over the course of its lengthy "year" (orbit around the Sun). This variation in solar energy is expected to cause the thickness of Pluto's atmosphere to change markedly in different parts of its orbit.

Charon is similar in size to Pluto. The diameter of Pluto is, approximately 2,284 kilometers, and Charon's is approximately 1,192 kilometers. The average distance between Pluto and Charon is 19,700 kilometers. Because of this close proximity, along with their similar sizes, Pluto and Charon are referred to as a double planet. Charon orbits Pluto in 6.39 days—the orbital period of Charon is the same as the rotation period of Pluto. As a result, Pluto always points the same face toward Charon. In fact, an observer on the surface of Pluto would always see Charon in the same position relative to the horizon.

The orbit of Charon is not in the plane of Pluto's orbit about the Sun. Instead, the plane swept out by Charon's orbit is almost perpendicular to the plane swept out by Pluto's orbit. When the plane of Charon's orbit presents

Pluto Compared with Earth

Parameter	Pluto	Earth
Mass (1024 kg)	0.0125	5.9742
Volume (1010 km3)	0.715	108.321
Equatorial radius (km)	1,195	6,378.1
Ellipticity (oblateness)	0.0000	0.00335
Mean density (kg/m3)	1,750	5,515
Surface gravity (m/s2)	0.58	9.80
Surface temperature (Celsius)	−238	−88 to +48
Satellites	3*	1
Mean distance from Sun		
millions of km (miles)	5,899 (3,657)	150 (93)
Rotational period (hrs)	−153.2928	23.93
Orbital period (days)	90,588	365.25

*In May, 2005, the discovery of two small Plutonian moons, Nix and Hydra, brought the total (with Charon) to three.
Source: National Space Science Data Center, NASA/Goddard Space Flight Center.

its edge to the Earth, a series of occultations and transits between Pluto and Charon occur. This series of transits and occultations results in a series of mutual eclipses being observed from the Earth. These mutual eclipses can be observed at two positions in Pluto's orbit, and so they occur every 124 years. Mutual eclipses last for about 6 years. A series of mutual eclipses that began in 1985 made possible the measurement of the sizes of Pluto and Charon reported above.

The surface temperature of Pluto is somewhat uncertain because the fraction of sunlight that it reflects (known as the albedo) is uncertain. Pluto's surface temperature ranges from 45 to 60 kelvins. The uncertainty arises because the surface composition of Pluto and the extent of its atmosphere are uncertain. The surface temperature of Charon is estimated to be between 8° and 10° warmer than Pluto's.

The density of the Pluto-Charon system has been determined to be of the order of 1,800 kilograms per cubic meter. This density, almost twice that of water, indicates that Pluto and Charon are composed of a variety of ices and that as much as half of their mass could be made up of rocky material. The surface of Pluto has, in fact, been

determined to contain methane ice. It is thought that, rather than the methane being uniformly distributed over the surface, there are two large polar ice caps made of methane and a thin, warmer equatorial region, where the methane has become depleted, leaving water ice.

Pluto is too small to trap a permanent atmosphere, but in the late 1980's, a thin atmosphere of methane was detected. Scientists believe that the atmosphere of Pluto was at its thickest during this period because Pluto was near its perihelion passage. At perihelion, its closest approach to the Sun, Pluto receives more energy from the Sun than it does during other parts of its orbit; in other words, it is heated more strongly. Methane is frozen under the conditions prevailing on Pluto. When it is heated sufficiently, it will form a gas directly from the solid without first forming a liquid, a process is called sublimation. It has been theorized that the atmosphere detected on Pluto may result from sublimation of methane from the equatorial region of its surface to form an atmosphere. Pluto will have to be observed through its entire orbit before it is known if it has an atmosphere throughout that orbit, and if it does, how the atmosphere's thickness specifically varies. It has been suggested that when Pluto is close to aphelion, only the side of Pluto facing the Sun would be warm enough to maintain a methane atmosphere. The atmosphere on the far side of Pluto would precipitate on its surface as frost.

The surface of Charon has been determined to be covered with water ice; no frozen methane has been detected. It is expected, however, that the interior of Charon contains methane. The composition of Charon is similar to that of some of the satellites of the Jovian planets. In fact, the surface of Charon appears to be almost identical in composition to that of Miranda, one of the satellites of Uranus. Charon is not expected to trap an atmosphere, even temporarily. It is difficult to make an exact determination, but an upper limit of no more than one-twentieth of the thickness of Pluto's atmosphere has been determined. Pluto's equatorial region is depleted in methane and thought to have the same composition as its similarly methane-depleted satellite, Charon.

Methods of Study

Most of the information currently available about the dual planet has been derived from the electronic recording of telescopic images of Pluto and then computer processing of these images. The rotation period of Pluto was measured by noting that the brightness of Pluto varies periodically with the rotation period of the planet. The brightness varies because the surface distribution of methane ice and water ice is not uniform, and different ices reflect different amounts of light.

Charon was discovered while James Christy examined some electronic images of Pluto. He noticed a bump on the edge of Pluto that appeared to move; this "bump" was Charon. No ground-based telescope was able to separate Pluto and Charon into well-resolved images, however, because of the Earth's atmosphere.

The atmospheres of Pluto and Charon have been studied by two different methods: occultations and spectroscopy. An occultation occurs when the light from an astronomical object is extinguished by another celestial object, such as when Pluto passes in front of a star. The observation of occultations is the standard technique used to determine whether a planet has an atmosphere or rings. If the planet has no atmosphere, it is possible to observe the star with undiminished brightness until the disk of the planet crosses it. It then disappears completely and reappears with its usual brightness. If a planet has an atmosphere, light from the star dims gradually as starlight passes through the atmosphere of the planet. When it reappears, it is faint and brightens as the planet moves farther away from the star. The atmosphere of Pluto was first detected in this manner.

Spectroscopy involves analysis of light reflected by Pluto. Different wavelengths are reflected by different degrees, and some are completely absent from the reflected light. The spectrum of reflected light can be used to identify chemical elements and compounds present on the surface of a planet and in its atmosphere. This procedure works because each element or compound produces a unique spectrum that can be measured in the laboratory. The infrared spectrum of Pluto has also been probed to add to the information. The main problem

The 1994 Hubble images of Pluto (small images, upper-left insets), have been enhanced in the larger images to reveal surface and atmospheric features. (NASA)

encountered in the Pluto-Charon system is that normally the spectra of Pluto and Charon are obtained simultaneously. Mutual eclipse events described above have enabled the spectrum of Pluto alone to be obtained when Charon is behind Pluto. This Pluto spectrum can then be subtracted from the usual combined spectra to obtain the spectrum of Charon. Using this method, scientists have been able to determine the different surface compositions of Pluto and Charon.

Occultations could also be used to measure the sizes of Pluto and Charon, but instead scientists have used the series of mutual eclipses. The rotation period of Charon about Pluto is known, so if the durations of the eclipses of Charon by Pluto are timed (and visa versa), these times can then be used to estimate the diameters of Pluto and Charon. Masses of the outer planets are usually measured by their effects on the orbits of planets closer to the Sun. This method, however, has not worked in the case of Pluto and Charon, because their combined mass is too small to have an observable effect on the next closest planet, Neptune. Fortunately, however, the discovery of Charon made it possible to determine the mass of Pluto from the orbital period of Charon. Kepler's third law of planetary motion states that the square of a planet's orbital period divided by the cube of its orbital radius is equal to a constant. The constant depends on the mass of the object orbited; hence, scientists have found that

the mass of Pluto is about one five-hundredth Earth's mass. The mass of Charon has been determined from its size by assuming it has the same density as Pluto.

Prior to Pluto's demotion in 2006 to dwarf planet status, it had often been said that it was the only planet yet to be explored by spacecraft. Indeed, due to the tremendous distance between Pluto and Earth, the best way to gain information about Pluto and Charon would be an instrumented spacecraft flying close to the illusive system or even perhaps entering long-term orbit about Pluto. A number of proposals were entertained to conduct the flyby, including one that would use a gravity assist from Jupiter in order to get to Pluto before Pluto's orbit took it so far from the Sun that its atmosphere would freeze and cover the surface. A Pluto flyby was approved and funded, then canceled, followed by a new proposal that in turn failed to come to fruition. Eventually, the New Horizons spacecraft was proposed, approved, designed, and then launched on January 19, 2006. The spacecraft left Earth behind with the fastest speed ever attained by a human-made object. It passed the Moon in a mere few hours and reached Jupiter in only one year. Nevertheless, New Horizons will not arrive in the Pluto-Charon system until 2015.

New Horizons, if successful, will provide a wealth of new information about the Pluto-Charon system. Scientific instruments incorporated into the spacecraft include a long-range reconnaissance imager, a near-infrared imaging spectrometer, an ultraviolet imaging spectrometer, an electrostatic analyzer, a time-of-flight ion and electron sensor, a radio science experiment, and a dust counter. Incredibly, the total mass of these instruments is a mere 31 kilograms, and they operate on only 21 watts of electrical power.

Context

Once considered the outermost planet, Pluto remains the most difficult to investigate. What has been learned about it indicates that it is different from all the other planets. The other four planets of the outer solar system—the "gas giants" Jupiter, Saturn, Uranus, Neptune—have low densities and are composed primarily of gases. The density of Pluto is greater, indicating the presence of some rocky material. Nevertheless, the density is lower than the densities of the terrestrial

Some Facts About Charon

Mean distance from Pluto (km)	19,600
Sidereal orbit period (days)	6.38725
Sidereal rotation period (days)	6.38725
Orbital inclination to Pluto (degrees)	0.0
Orbital eccentricity	0.0
Equatorial radius (km)	593
Mass (10^{21} kg)	1.62
Mean density (kg/m^3)	1,850
Surface gravity (m/s^2)	0.31
Escape velocity (km/s)	0.60
Albedo	0.38
Apparent visual magnitude	16.8

Source: Data are from the National Aeronautics and Space Administration/ Goddard Space Flight Center, National Space Science Data Center.

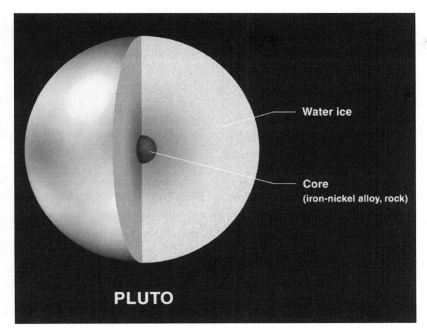

Pluto's projected composition; surface methane ice sublimates, creating a thin methane atmosphere, when the dwarf planet is facing the Sun. (Lunar and Planetary Institute)

planets of the inner solar system (Mercury, Venus, Earth, and Mars). Pluto has features in common with the Galilean satellites of Jupiter and some of the satellites of Saturn, Uranus, and Neptune, but none has exactly the same makeup. Charon is in many ways similar to an asteroid. In fact, models developed for asteroids covered with water ice are applicable to Charon, and they have been used profitably in an attempt to gain a deeper understanding of Charon.

This pair of small worlds may seem insignificant in comparison with the other, larger planets of the outer solar system. However, if scientists are ever to develop a complete understanding of the origins and evolution of the solar system, they will need detailed knowledge of all its members. Pluto is usually the farthest planet from the Sun, but in many ways it has more in common with comets, meteoroids, and asteroids than it does with the other major planets. Pluto could well be one of a large number of similar objects in the outer solar system. Although there are no large objects in the solar system even ten times farther away from the Sun than is Pluto, there is a spherical swarm of comets far from the Sun, called the Oort Cloud. In developing an understanding of Pluto and Charon, scientists may be laying the groundwork for a better understanding of the Oort Cloud.

Study of the Pluto-Charon system may also answer some questions about Neptune and its satellites. One of the theories for the origin of Pluto is that it was once a satellite of Neptune. That appears less likely since scientists have learned that Pluto has its own satellite. However, an understanding of its eccentric orbit would provide a definitive answer. The mysteries of the solar system include the eccentric orbit of Neptune's satellite Nereid and the clockwise direction (as viewed from the north pole) of the orbit of another satellite, Triton, when all the other large satellites of the solar system orbit their planets in a counterclockwise direction. Both of these oddities could be explained by a collision in which Pluto broke free. However, that theory lost favor with the discovery of Quaoar, Sedna, and Eris and the realization that the Kuiper Belt is likely populated with almost countless bodies of lesser size than these three icy objects. The common belief presently held is that Pluto and Charon were created in the early Kuiper Belt by a giant impact, not unlike the impact theory for the formation of the Earth and its Moon. More research is needed to pinpoint the origin of this unusual dual-object system in the outer solar system.

It was in part for these reasons that the International Astronomical Union (IAU) moved to reclassify Pluto as a dwarf planet even though it still met many of the restated criteria for classification as a planet. That definition now requires that a body orbit about the Sun, that it is in hydrostatic equilibrium, and that it has cleared out its environment. Pluto meets only the first two of those criteria. In addition, since 2000 three objects roughly the same size as Pluto have been discovered well beyond Pluto's orbit. These bodies (Quaoar, Sedna, and Eris) are believed to be Kuiper Belt objects, and Pluto is likely the first of the Kuiper Belt objects to have been discovered. As such, it is the model for the rest of what may be a huge class of similar objects at ever-increasing distance. Some have coined the term "plutinos" to describe icy bodies coming from the

Clyde Tombaugh: Searching for Planet X

(National Archives)

In many ways, Clyde Tombaugh was like his astronomer hero William Herschel, who discovered Uranus unexpectedly during a routine sky survey in 1781. Both were dedicated amateur astronomers and skilled telescope makers who devoted hours to tedious observations. Tombaugh, however, was only twenty-four years old when he discovered Pluto, while Herschel was in his early forties. Furthermore, Tombaugh's yearlong search for Planet X lasted much longer than that of either Herschel or Johann Galle, who discovered Neptune in 1846 on the first night he looked for it, lying less than 1 degree from its predicted position.

The search for Pluto was complicated by the fact that its orbit is highly eccentric— sometimes even passing inside the orbit of Neptune—and has a large inclination of about 17 degrees from the mean plane of the other planets. It is now known that Percival Lowell's predictions for the position of Pluto were based on faulty calculations, and its discovery within 6 degrees of the predicted location was only a coincidence. Fortunately, Tombaugh did not limit his observations to the predicted area of the sky or to the region close to the mean orbital plane of the planets.

When James Christy discovered Pluto's moon, Charon, in 1978, it was conclusively demonstrated that the mass of Pluto was far too small to cause observable deviations in the orbits of Uranus and Neptune; thus the two larger planets' orbits could not be used to predict Pluto's position. In the 1990's, several icy objects much smaller than Pluto were discovered just beyond its orbit in the Kuiper comet belt with periods of about 300 years, compared with Pluto's 248-year period.

Kuiper Belt. Because of the new classification scheme adopted by the IAU in 2006, the solar system officially consists of eight planets, at least three dwarf planets (Ceres, Pluto, and Eris), and numerous satellites, asteroids, comets, and minor bodies.

In an attempt to clarify its controversial planetary classification and minimize the worldwide discontent expressed over Pluto's elimination from full-fledged planet status, an executive meeting of the International Astronomical Union (IAU) held in June, 2008, in Oslo, Norway, proposed defining a plutoid as a solar system body beyond the orbit of Neptune having enough mass to assume a nearly spherical shape but not able to clean its orbit of other material, as the eight planets have. Pluto would therefore be the first plutoid object. Eris would be the second. Under this definition, Ceres would remain a dwarf planet but could not be considered to be a plutoid, since it exists in the asteroid belt between Mars and Jupiter. For many, this proposal did little to mitigate earlier objections. This classification scheme left the solar system with only the one dwarf plantet, which until 2006 most astronomers had been quite content to consider to be the largest asteroid. More plutoids were expected to be found, located even farther out in the Kuiper Belt beyond Eris, as technology permitted their detection. In the

period following the IAU's adoption of the reclassification scheme, there remained much controversy, not only among professional astronomers but also among amateur astronomers, teachers, and schoolchildren.

Whether or not Pluto is a planet, its further investigation is hindered until the New Horizons spacecraft can fly through the Pluto-Charon system in 2015. Clyde Tombaugh provided the first evidence of Pluto's existence. Some proposed that a spacecraft be sent to explore Pluto at close range, and that it be christened Tombaugh. Clyde Tombaugh passed away in 1997 before such a flyby mission was approved and launched. However, the New Horizons probe presently is on track for a 2015 encounter with Pluto and Charon. On board that spacecraft are some of Tombaugh's ashes. A flyby of Pluto would complete humanity's first exploration of what once was considered a nine-planet solar system.

Stephen R. Addison

Further Reading

De Pater, Imke, and Jack J. Lissauer. *Planetary Sciences.* New York: Cambridge University Press, 2001. A challenging and thorough text for students of planetary geology. An excellent reference for the most serious reader with a strong science background.

Faure, Gunter, and Teresa M. Mensing. *Introduction to Planetary Science: The Geological Perspective*. New York: Springer, 2007. Designed for college students majoring in Earth sciences, this textbook applies general geological principles to bodies throughout the solar system. Excellent for learning comparative planetology.

Hartmann, William K. *Moons and Planets*. 5th ed. Belmont, Calif.: Thomson Brooks/Cole, 2005. An updated version of a classic text that covers all aspects of planetary science. Provides a full description of our contemporary understanding of Pluto.

Levy, David. *Clyde Tombaugh, Discoverer of Planet Pluto*. New York: Sky, 2007. Written by an amateur comet hunter (of Shoemaker-Levy fame), this book details the detective story that was Clyde Tombaugh's quest at the Lowell Observatory in Flagstaff, Arizona, to find Planet X, which we now call Pluto.

Littmann, Mark. *Planets Beyond: Discovering the Outer Solar System*. New York: Dover, 2004. A history of exploration of the outer solar system as well as of Earth-based observations of Uranus, Neptune, and Pluto. Essays are interjected into the text profiling astronomers who have made important contributions.

Morrison, David, and Tobias Owen. *The Planetary System*. 3d ed. San Francisco: Pearson/Addison-Wesley, 2003. An introduction to the properties of all the major objects of the solar system, at the level of a college distribution course in astronomy. A good source of data about the solar system, taking a comparative planetology approach.

Tombaugh, Clyde W., and Patrick Moore. *Out of the Darkness: The Planet Pluto*. New York: New American Library, 1981. A classic about the discovery of Pluto by its discoverer, a well-known popularizer of astronomy. Tombaugh describes his career as well, and coauthor Moore describes the discovery of outer planets Neptune and Uranus. Of interest primarily for its historical perspective, since much of the material on the physical properties of Pluto and Charon is outdated.

Tyson, Neil deGrasse. *The Pluto Files*. New York: W. W. Norton, 2009. When the International Astronomical Union devised a new scheme for classifying solar-system objects, Pluto was suddenly demoted to dwarf-planet status, much to the displeasure of many astronomy buffs and large segments of the astronomical community. This book provides an account of Pluto's various classifications, from ninth planet to its loss of planetary status.

SATELLITES

Category: Space Exploration and Flight

Artificial satellites permanently stationed in space perform many important economic, military, and scientific missions. Uncrewed artificial satellites are the chief instrument of space exploration and provide the only means of obtaining permanent utilization of space.

Overview

Virtually all objects in space are satellites of one body or another. Satellites range in size from galaxies such as the Large and Small Magellanic clouds in orbit about the Milky Way to microscopic flakes of paint in low-Earth orbit that have eroded from artificial spacecraft. In practice, the word satellite is reserved for uncrewed spacecraft in Earth orbit. Crewed spacecraft are usually referred to individually by name, such as the International Space Station. Nonfunctional objects of artificial origin are regarded as orbital debris. Natural satellites of stars are more properly referred to as planets, while natural satellites of planets are more properly referred to as moons.

Satellites travel on elliptical trajectories called orbits, which are freely falling paths determined by the local gravitational field. Although satellites are indeed falling, they are also traveling sideways at extremely high speeds, on the order of 7 kilometers per second (5 miles per second) at 200 kilometers altitude (130 miles). The combination of free fall and high lateral velocity creates a closed trajectory that carries the satellite around Earth repeatedly.

The point on the orbit nearest to the earth is called the perigee; it is also the point at which the satellite has the greatest velocity. The point farthest away is called the apogee. That is also where the satellite velocity is least. If space were a perfect vacuum, satellites would orbit forever, but the atmosphere has no distinct end, and gradually fades away with altitude. Satellites orbiting at altitudes from 200 to 600 kilometers (130 to 400 miles) encounter enough residual atmosphere to create significant aerodynamic drag. Over the months, these low-Earth-orbit satellites lose energy and decrease in apogee until the apogee equals the perigee and the orbit is a circle. The satellites then drop closer to Earth on a spiral path, accelerating as they do so. Eventually, they enter regions where the atmosphere is too thick for them to continue in orbit. Aerodynamic drag becomes so strong that all of the satellite's energy is converted into heat in a matter of minutes.

The air around the satellite becomes hot enough to glow, and exposure to the heat burns up the satellite.

Satellites orbiting below 200 kilometers (130 miles) reenter Earth's atmosphere in a matter of months. Those orbiting above 600 kilometers (400 miles) seldom reenter.

Satellites are classified according to user (commercial, military, or scientific) and according to mission (communications, remote sensing, or experimentation and measurement). Commercial satellites belong to private businesses. Military satellites support military operations. Scientific satellites perform experiments or make measurements in support of scientific research.

A satellite is only one part of a space mission's architecture, an assembly which consists of the satellite, the launch system necessary to place it in orbit, the ground support system necessary to control the satellite and communicate with it, and a data analysis and information management system to exploit the data gathered by the satellite.

The Satellite Design Process

The satellite design process begins with the delineation of the satellite mission. A mission to photograph Earth from space, for example, might be expressed in terms of the goal that all areas of Earth between 45 degrees north latitude and 45 degrees south latitude be photographed with sufficient clarity that objects as small as 10 meters across can be imaged clearly. This requirement immediately eliminates all orbits of less than 45 degrees inclination and makes the orbital altitude of the satellite heavily dependent on camera quality: high-resolution cameras will be able to fulfill the requirement from greater altitudes than low-resolution cameras. In this way, the mission is expressed in the form of a set of requirements for orbital altitude, inclination, life span, launch date, and other needs which the satellite must fit.

A satellite is composed of the payload and the support bus. The payload consists of those components which perform the primary mission of the satellite. Component choice is driven by the best fit of available hardware to mission requirements. The components chosen will in turn determine payload parameters such as mass and volume, and payload demands such as power consumption, data storage and transmission, and attitude control.

The bus contains various systems to support the payload and provide electric power, thermal control, attitude control and propulsion, communications, and structural support. Bus components must be chosen that are capable of filling all of the payload demands as well as supporting the bus itself.

Total mass and volume are determined once payload and bus design are complete. Total mass and volume together with orbit requirements determine the choice of launch vehicle.

No satellite design process is complete without the development of ground sites. Ground sites monitor the status of the satellite and issue commands as necessary to maintain proper function or to correct anomalies in function. Ground sites receive data sent down by the satellite, process the data into a form intelligible to the

Antennas and satellite dishes generally have a parabolic shape and are used to receive satellite signals on Earth. (Photos.com)

user, and deliver it. Ground site personnel continually track the satellite, noting inevitable changes in orbit and issuing predictions for future passes within range of the ground site.

Power

The power system provides the electric power needed to operate electrical and electronic components. Solar cells are usually the primary source of power, converting sunlight to electricity. What is not immediately required for satellite operations is stored in rechargeable batteries for later use. The power requirements of the payload and bus together determine the size of power system components. Solar cells must have enough area to collect all the power needed by the satellite plus more to provide a margin of safety. Because solar cells degrade over time in the harsh space environment, they must be built larger than initially required to guarantee that enough capability remains after years of degradation to continue operating the satellite. The number and size of batteries must be sufficient to meet the voltage and current demands of the payload and bus.

Power consumption must be carefully managed on board satellites. Consumption of electricity inevitably generates heat, which cannot easily escape in the vacuum of space and becomes a challenge for the thermal control system. Batteries build up internal pressure when charging and are in danger of bursting and destroying the satellite if overcharged. On the other hand, batteries that discharge too deeply are in danger of dying completely. Also, electronic components that lose power or receive too little voltage (an undervoltage condition) may cease operating or undergo an uncommanded reset when normal conditions return. Power system conditions such as voltage, current, and temperature are monitored at critical locations with the results transmitted to satellite operators on the ground.

Thermal Control

The thermal control system maintains proper temperature throughout the satellite. It removes heat from components in danger of overheating from electric power consumption or exposure to the Sun, and provides heat to components in danger of freezing from exposure to the cold vacuum.

Attitude Control and Propulsion

The attitude control system maintains the satellite in the proper orientation required for the satellite to fulfill its mission. Communications satellites must have antennas permanently pointed toward Earth's surface, for example, while the Hubble Space Telescope must be constantly looking at the object being photographed.

The simplest type of attitude control system is none at all; the satellite is allowed to tumble uncontrollably. This requires the use of antennas that broadcast in all directions at once, so that communication with the ground is never interrupted. This also means that most of the broadcast power is wasted on transmissions into empty space and that only a small fraction of the power reaches the ground. This is acceptable only for the simplest types of low-Earth-orbit satellites.

Oblong satellites can be oriented so that the long axis points toward Earth and couples to tidal gravitational forces to provide gravity gradient stabilization. Once gravity gradient stabilization is achieved, the satellite will permanently present one face toward Earth, where cameras, remote sensing instruments, and communications antennas may be advantageously mounted. Gravity gradient stabilization is usually achieved by building a telescoping boom into the satellite structure, which deploys when the proper orientation is obtained. When extended, the end of the boom closest to Earth feels the strongest gravitational field and is continually pulled downward. That continuous downward pull keeps that end pointed toward Earth.

Active attitude control systems include momentum wheels and control moment gyroscopes. Momentum wheels are spun up in one direction so that the satellite will spin in the opposite direction in reaction. Three momentum wheels mounted in three perpendicular directions provide attitude control about any rotation axis. When the spin axis of a control moment gyroscope is altered, complicated reaction forces are created that may be used to rotate the spacecraft. Both of these systems have the virtue of reorienting the satellite without consuming propellant.

Active attitude control requires the satellite to have some knowledge of its orientation with respect to the outside world. The location of the Sun can be determined through the use of sensors that respond to visible light to indicate which side of the spacecraft is facing the Sun and which is in shade. Earth sensors respond to infrared radiation from the comparatively warm Earth. Star sensors look for the light from very bright stars. Stable platforms controlled by gyroscopes maintain a constant orientation regardless of the rotation of the spacecraft.

Communications

The communications system keeps the satellite in contact with the ground support system and moves data and commands to and from the satellite. The communications system includes transmitters and receivers, data encoders and decoders, data storage and retrieval elements (memory), and antennas. High-gain directional antennas carry the maximum amount of data with the minimum amount of power, but must be accurately pointed toward the reception site. This requires additional equipment to control the pointing of the antenna and maintain communications lock. The antenna may move itself, or the attitude control system may be tasked to reorient the entire satellite.

Orbital speeds of the order of 7 kilometers per second (5 miles per second) create significant shifts in the frequency of radio waves transmitted or received by satellites. Frequency goes up as the satellite approaches a ground site and falls as the satellite recedes, a phenomenon known as the Doppler shift. The ground site must continuously adjust frequency of both transmission and reception so that communication is continuous and no information is lost.

Most satellites are in range of a ground site for only ten minutes or less at a time and only during the infrequent occasions when their orbit takes them over the ground site location. Data collected at other times must be stored on board for relay to the ground during the next pass.

Structure

The structural system holds the parts of the satellite together and protects the components of the satellite from the high accelerations and intense vibrations experienced during launch. Structures range from simple frames to hold the components of the satellite in place to complicated mechanical systems folded and stowed during launch that must unfold and extend instruments upon deployment. The structure must not respond resonantly to vibrations generated by the launch vehicle or the satellite will shake itself to destruction. Special composite materials and honeycomb construction keep structural members lightweight without sacrificing strength.

Satellite Construction and Testing

The high costs of launch and the inability to make repairs on malfunctioning satellites demand high reliability and long operational lifetimes. Both are expensive and difficult to achieve. Altogether, these requirements force satellite designers and builders to make every attempt to make the satellite perfect the first time and every time. Components are extensively tested individually, and each system is tested and retested as new components are added. Complete systems are tested individually, and then tested and retested as they are linked into the final satellite assembly. Finally, the complete satellite is tested and retested under conditions simulating spaceflight as closely as possible.

The quest for perfection begins at the component level. Items for use in satellites must meet rigorous requirements. Materials cannot emit water vapor or volatile organic compounds in a vacuum. They must not chemically break down, degrade, or darken under exposure to ultraviolet light or atomic oxygen. Electronic parts and components must not be susceptible to ionizing radiation. Electrical systems must not be susceptible to the build-up and discharge of static electricity.

Complete satellite assemblies must survive a harsh launch environment. Launch vehicle accelerations can produce the equivalent of eight to ten times normal weight in the satellite. Rocket exhaust plumes generate strong vibrations and intense noise that can vibrate poorly constructed assemblies to destruction. Satellites therefore undergo vibration testing on massive shake tables that realistically simulate the launch vibration environment. After vibration testing, the satellite is placed in a vacuum chamber and run through heating and cooling cycles that mimic what the satellite will encounter in space.

All stages of satellite construction are extensively documented. Even after all this testing, satellites fail on orbit. Since a failed satellite cannot be retrieved for study, the only way to analyze what went wrong is to review the documentation and deduce the cause of the failure. A complete and thorough record of the design and construction process is essential.

Tracking Satellites

The U.S. Space Command (USSPACECOM) catalogs and tracks every object in Earth orbit greater than 10 centimeters (4 inches) in length with ground-based radar and electro-optically enhanced telescopes. Continuous space surveillance allows U.S. Space Command to predict when and where a decaying space object will reenter Earth's atmosphere in order to prevent an innocent satellite or inert piece of debris from triggering missile-attack warning sensors of the United States or other countries upon reentry. It also charts the present position and anticipated motion of space objects, detects new manmade objects in space, and determines their country of origin. An extremely important function of space surveillance is to inform the National Aeronautics and Space Administration (NASA) of the identity and path of objects that may endanger the space shuttle.

End-of-Life Operations

Space is becoming crowded. The Soviet Union launched Sputnik 1, the first artificial Earth satellite, in October,

1957. The United States launched its first satellite, Explorer 1, in January, 1958. Both have long since decayed and burned during reentry. The oldest satellite still in orbit is Vanguard 1, launched in March, 1958. As of June 6, 2001, U.S. Space Command reported 2,728 satellites in orbit, while 2,569 other satellites had undergone orbital decay and burned on reentry since 1957.

Satellites still in orbit degrade in the harsh space environment, shedding small particles of debris, such as paint flecks and pieces of thermal blanket. In extreme cases, old satellites are completely destroyed when aging batteries burst or leftover propellant spontaneously explodes. As of June 6, 2001, U.S. Space Command reported 6,150 pieces of debris in orbit that were 10 centimeters or greater in length. Satellites in low-Earth orbit run a significant risk of collision with a piece of orbiting debris. At collision velocities on the order of 10 kilometers per second (about 7 miles per second) even a tiny fleck of paint can do significant damage.

In an effort to slow the rate at which new debris is being created, satellite designers routinely include end-of-life planning in the satellite design process. At end-of-life, batteries are disconnected from solar panels to prevent destructive overcharging, and any unused pressurized liquids or gases are vented into the vacuum. The last few gallons (or pounds) of propellant are consumed in an orbital adjustment burn which either forces low-Earth-orbit satellites to reenter the atmosphere and burn up, or moves higher-altitude satellites to disposal orbits where they do not present a hazard to other spacecraft.

Observing Satellites

Satellites shine by reflected light and are visible to the naked eye for a short time just before sunrise and just after sunset. During these periods, the background sky is dark enough for dim objects to be seen by observers on the ground, but satellites passing overhead are still illuminated by the sun. There are so many satellites in orbit that every morning and evening, several pass over virtually every location on Earth. Satellites of the Iridium group of communications satellites have large, highly polished solar panels that can be extremely bright when the sun is reflected in them. Sightings of so-called Iridium flares are extremely common.

Billy R. Smith, Jr.

Further Reading

Heavens Above. (www.heavens-above.com) Provides easy-to-use information about satellite passes, both morning and evening, for almost every location on Earth. The user inputs either a place name or latitude and longitude information, and the Web page returns pass predictions for all visible satellites for the coming days. Star maps showing the start, stop, and path of the pass are also available. High-visibility objects, such as the International Space Station and Iridium flares, are specifically noted.

Maral, Gerald, and Michel Bousquet. *Satellite Communications Systems: Systems, Techniques, and Technology*. 3d ed. New York: John Wiley & Sons, 1998. Offers a detailed analysis of satellite communication system construction and operation.

Montenbruck, Oliver, and Eberhard Gill. *Satellite Orbits: Models, Methods, Applications*. New York: Springer Verlag, 2000. A textbook on orbital mechanics covering all aspects of satellite orbit prediction and determination.

U.S. Space Command. (www.peterson.af.mil/usspace/index.htm) The U.S. Space Command Web site provides links to the current satellite box score and satellite space catalog.

SPACE DEBRIS

Category: Atmosphere and air pollution

Space debris—which consists of nonfunctioning spacecraft, rocket bodies, refuse from missions, and fragments thereof—poses a hazard for space missions, satellite-based services, and people both in space and on earth. As space-faring nations have become more aware of the dangers of this debris, they have worked to minimize its generation during operations in space.

Overview

Since 1957 human beings have launched thousands of satellites and other spacecraft. Most of the spacecraft launched successfully achieve orbit. Those that explode after attaining orbit altitude and those that fail after achieving orbit become space debris (also known as orbital debris or space junk). Anything that reaches orbit altitude—about 300 kilometers (186 miles) above the earth's surface—becomes a satellite of the earth. Once in orbit, objects are constantly under the pull of the earth's gravity, and, in time, they slowly fall from orbit. The greater the distance from the earth, the longer an object will remain in orbit. Above 1,000 kilometers (621 miles) objects can remain in orbit for at least a century, objects orbiting at an altitude of 800 kilometers (497 miles) are likely to fall to earth within decades, and those at altitudes between 200 and 600 kilometers (124 and 373 miles) tend to remain in orbit for several years at best.

The U.S. Air Force Space Surveillance Network, which routinely tracks artificial objects orbiting the earth, has cataloged roughly 19,000 debris objects larger than 10 centimeters (4 inches) in diameter. An estimated 500,000 orbiting particles are between 1 and 10 centimeters (0.4 and 4 inches) in diameter. Particles measuring less than 1 centimeter in diameter probably number in the tens of millions. Most of the debris orbits within 2,000 kilometers (1,243 miles) of the earth's surface, with the greatest concentrations accumulating at altitudes between 800 and 850 kilometers (497 and 528 miles).

Each object in orbit runs the risk of running into another object. The volume of space surrounding the earth is immense, and the chances of a collision between two objects are relatively low; however, the likelihood of collision increases when the objects occupy the same orbit. Because certain orbits are particularly desirable for satellites used for communications and surveillance purposes, various nations and commercial interests place their satellites into these positions, thereby increasing the chances of collision.

Many different kinds of space debris orbit the earth. From the 1960's through the mid-1980's, nations deliberately destroyed orbiting satellites while testing weapons for antisatellite warfare. Other forms of space debris have less dramatic origins, such as astronaut Ed White's glove, which slowly drifted away from his Gemini spacecraft in 1965. Each item adds to the ever-increasing number of human-made objects orbiting the earth. Collisions between objects, and explosions of residual fuels in abandoned rocket engines, break existing debris into many smaller pieces.

Objects ranging in size from spent rocket boosters and nonfunctional satellites to small chips of paint, solid-fuel fragments, and coolant droplets have the potential to damage spacecraft. It is not merely the mass of an object that poses a danger but also its high velocity. At orbits below 2,000 kilometers, debris travels at speeds of 7 to 8 kilometers (4.3 to 5 miles) per second, so that even tiny particles can pit space shuttle cockpit windows and damage unshielded satellite components.

Hazards

The space debris population has grown great enough that it has become standard practice to shift unmanned satellites

---

Naturally Occurring Space Material

The term "space debris" is sometimes used to refer to naturally occurring material as well as that generated by human activity. However, as space technology consultant Mark Williamson notes in *Space: The Fragile Frontier* (2006), natural space objects such as meteoroids, meteorites, and cosmic dust are inherent to the space environment, unlike anthropogenic material. Humankind can guard against the flux of natural space material but cannot halt or lessen it; humans can increase it, however, by reducing a single larger body into countless smaller fragments.

An estimated 25 million bits of natural space material collide with the earth each day, the majority of it in the form of cosmic dust. These particles are so small that they do not even appear as meteors as they pass into the atmosphere. Most meteors that are seen are particles the size of a pea, and sometimes the larger ones reach the earth's surface as meteorites.

Occasionally as asteroid-sized objects or comet collides with the earth. Such collisions have global implications. The impact destroys the comet and forms a huge crater. An enormous amount of gas and dust is carried into the atmosphere, creating a blanket of debris that blocks the sunlight. This begins as "nuclear winter"-type effect, which can last anywhere from a few months to several years. During this time most life-forms will die as a result of the disruption of the food chain. Many scientists believe that such an event led to the extinctions of dinosaurs about 65 million years ago.

Although giant impacts are one cause of "nuclear winter," cosmic dust can produce the same effect. The solar system periodically runs into a cosmic dust cloud, thereby dramatically increasing the amount of dust that enters the atmosphere. A similar situation also results from periodic meteor storms.

As the earth runs into the debris of old comets, the number of meteors that enter the atmosphere increases. The earth occasionally encounters a particularly dense region of comet debris. During such meteor storms, thousands of meteors can be seen each hour. The most notable is the Leonid meteor storm, which occurs every thirty-three years. Many scientists fear that increases in comet debris could knock out hundreds of satellites as they are hit by microscopic particles and greatly affect global positioning and communication capabilities.

out of harm's way when large debris (objects larger than 10 centimeters) is detected. Space shuttle flights have to adjust course to avoid debris reported by the Space Surveillance Network. The International Space Station is heavily shielded against objects smaller than 1 centimeter, but it has the capability to maneuver away from larger tracked objects.

Only one collision between large, intact satellites has ever occurred. In February, 2009, an operational U.S. Iridium communications satellite accidentally struck a deactivated Russian Cosmos communications satellite. Both spacecraft were destroyed, and more than 1,500 large fragments were generated. The amount of large debris had already been dramatically increased two years earlier, when in January, 2007, China conducted an antiweapons test in which it used its aging Fengyun-1C weather satellite as a target. The resulting destruction created roughly 2,600 large debris fragments and hundreds of thousands of smaller particles.

Efforts to minimize the problems associated with space debris include the boosting of geostationary satellites that have ended their missions out of their orbits (near 36,000 kilometers, or 22,369 miles, above the earth's surface) into a higher "disposal orbit." Similarly, deactivated satellites that operated at lower altitudes may be moved to even lower orbits that will decay more quickly, hastening the satellites' fall to earth. If a satellite fails to burn up in the atmosphere, however, it can present a threat to people and property on the earth's surface.

In 1978 a Soviet satellite with a nuclear power source survived reentry and strewed small amounts of radioactive material across Canada. The following year, large pieces of the Skylab space station withstood a fiery plunge through the atmosphere and scattered debris across western Australia. In 2001 a rocket upper stage that had been part of a 1993 global positioning satellite launch fell to earth in the Saudi Arabian desert. All of these incidents would have caused considerable damage if the debris had not landed in sparsely populated areas. Only one instance has been recorded of a person being struck by space debris: In 1997 a bit of woven metallic material from a Delta II rocket fuel tank hit an Oklahoma woman on the shoulder but did not injure her. On average, one piece of cataloged space debris falls out of orbit every day, usually burning up in the atmosphere.

Mitigation Measures

Careful design and operational measures can keep new space missions from contributing unnecessarily to the proliferation of space debris. For example, upper stages of launch vehicles can be placed at lower altitudes so that their orbits decay sooner. Since 1988 the United States has had an official policy of minimizing debris from governmental and nongovernmental operations in space, and the U.S. government approved a set of standard practices for the mitigation of space debris in 2001. The governments of France, the European Union, Japan, and Russia also have issued guidelines pertaining to space debris. Additional guidelines have been published by the United Nations Committee on the Peaceful Uses of Outer Space (COPUOS) and the Inter-Agency Space Debris Coordination Committee (IADC), a group established by the world's leading space agencies in 1993.

Cleaning up existing space debris remains an expensive and technologically challenging prospect. Proposed solutions have included hastening objects' fall to earth by using lasers to slow their orbits and conducting special robotic space missions to grab and haul debris. Solutions that are both technically feasible and economically viable have yet to be developed. In addition, the development of technologies for the cleanup of space debris is controversial because any methods capable of moving spacecraft have potential weapons applications.

A 2006 study sponsored by the U.S. National Aeronautics and Space Administration (NASA) Orbital Debris Program concluded that, if no new launches were conducted and no new objects introduced to earth's orbit, the number of objects falling out of orbit over the next half century would balance the number of new objects created through collisions. After 2055, however, the increasing number of collision-generated fragments—which would go on to create their own catastrophic collisions—would overtake the number lost through decaying orbits.

The international space community's concern in the wake of the 2006 NASA study, China's 2007 weapons test, and the 2009 satellite collision led to the first International Conference on Orbital Debris Removal, convened in December, 2009. Participants examined the many technical, economic, legal, and policy issues surrounding near-earth space cleanup, but they reached no conclusions regarding exactly how humankind might best address the worsening problem of space debris.

Paul P. Sipiera
Updated by Karen N. Kähler

Further Reading

Inter-Agency Space Debris Coordination Committee. *IADC Space Debris Mitigation Guidelines*. Vienna: United Nations, 2002.

Johnson, Nicholas L., and Darren S. McKnight. *Artificial Space Debris*. Updated ed. Malabar, Fla.: Krieger, 1991.

Klinkrad, Heiner. *Space Debris: Models and Risk Analysis*. New York: Springer, 2006.

National Research Council. *Orbital Debris: A Technical Assessment*. Washington, D.C.: National Academy Press, 1995.

Simpson, John A., ed. *Preservation of Near-Earth Space for Future Generations*. 1994. Reprint. New York: Cambridge University Press, 2006.

Smirnov, Nickolay N., ed. *Space Debris: Hazard Evaluation and Mitigation*. London: Taylor & Francis, 2002.

United Nations Committee on the Peaceful Uses of Outer Space. *Technical Report on Space Debris*. New York: United Nations, 1999.

Williamson, Mark. *Space: The Fragile Frontier*. Reston, Va.: American Institute of Aeronautics and Astronautics, 2006.

SPACE RESOURCES

Category: Ecological resources

The vastness beyond Earth's thin atmosphere is rich in extraterrestrial resources. Microgravity technologies have been developed to take advantage of those resources. Solar energy captured on Earth or in space is used to generate electrical power. Applications in communications, global monitoring, and the Global Positioning System have been developed to improve the quality of human life. Satellites document planetary biosphere changes that occur naturally or from human activity.

Overview

Launching Sputnik 1 in 1957, the Soviet Union began a race to develop technology that provided routine access to space. The region from low Earth orbit (LEO) outward to geostationary Earth orbit (GEO) is concentrated with satellites that peer regularly at Earth or with telescopes looking outward. The region between LEO and GEO is the most utilized with regard to space resources. There, some resources have present commercial profitability. Space beyond GEO remains largely for scientific exploration and resource speculation.

Communication Satellites

An object in GEO revolves about Earth's center in exactly one day. This means that as Earth rotates on its axis, a GEO object appears to hang directly overhead. Geostationary position is 35,800 kilometers above Earth's surface.

Early communications satellites were only put in LEO. The next push was to install operational systems in GEO to relay television images, data, and telephone signals around the world. LEO satellites have regained a share of communications traffic. These cross an observer's sky in ten to twenty minutes, so a constellation of satellites is required for continuous reception. Because LEO satellites are only 500 to 1,400 kilometers above Earth's surface, they can be reached with a signal much less powerful than one required for a geostationary satellite. Consequently, ground stations that provide uplinks and downlinks for LEO satellites can be modest. Thus, LEO satellites can provide portable telephone service and data links to underdeveloped areas.

Weather and Climate Observing Satellites

Geostationary weather satellites provide images in visible and infrared light. Polar-orbiting weather satellites survey virtually the whole Earth. Satellite instruments monitor stratospheric ozone concentrations, atmospheric particulates, temperature profiles as a function of atmospheric altitude, and pollutant levels (such as chlorofluorocarbons). Some measure surface, lower atmospheric, and ocean temperature variations to monitor suspected global warming. A great advantage of global weather systems has been advanced warning of hurricanes, tornadoes, and other destructive systems, resulting in tremendous savings of human life.

Navigation Satellites

The Global Positioning System (GPS) is a network of twenty-four Navstar satellites maintained by the U.S. Department of Defense. A person using a special receiver and security codes can determine a location to fewer than 18 meters. Without those security codes, accuracy is limited to around 100 meters. This is sufficient for civilians to drive to a location in a strange city or to navigate a ship. The military uses GPS not only for navigation but also to provide flight-path corrections to deployed smart weapons. GPS was incorporated into guidance and navigation systems aboard the space shuttle. GPS became a staple for search-and-rescue services and provides a means for detailed documentation of surface locations for commercial and scientific purposes.

In 2009, a GPS satellite launched by a Delta II booster lifted off from Cape Canaveral. At that point, the Delta II rocket had launched forty-seven GPS satellites in twenty years with only one failure. Since initial GPS deployment, various generations of GPS have been launched by Delta II, Atlas II, and Titan IV rockets. With this 2009 launch, there were thirty operational satellites, well beyond the minimum of twenty-four needed for the orbital constellation.

Reconnaissance, Remote Sensing, and High-Resolution Imaging Satellites

In 1960, the Russians shot down an American U-2 spy plane flying over Soviet territory. This incident underscored the military's desire to obtain high-resolution images in a less vulnerable manner. Soon, spy satellites, from the original Corona (cover name Discoverer) reconnaissance satellites that proved the utility of military intelligence gathering from orbit to modern classified electronic listening and imaging platforms took over from spy planes. Afterward, relying on assets from orbit became a major part of American military space programs. Resolution and other capabilities of military systems, naturally, remain secret.

The orbital vantage point not only is useful for reconnaissance and intelligence gathering but also provides a platform from which to perform Earth resources investigations. The story has been told of a Gulf of Mexico fisherman who, when shown an image taken from NASA's Skylab Earth Resources Experiment Package (EREP), stated that he had learned more about where to find rich schools of fish and where currents and abundant nutrients flowed within his patrol area than he had during a lifetime of working on the sea. Multispectral imaging could be used to conduct environmental studies as well as uncover a wide range of natural resources. From early astronauts using simple cameras to the Skylab EREP package, the concept of remote sensing was proven quickly.

Public access to satellite images began in 1972 with the Landsat satellite series. A similar program to observe the oceans, called Seasat, was developed with less success than Landsat. Early Landsat images had a resolution of 80 meters. During the Carter administration, NASA transferred Landsat operations to the National Oceanographic and Atmospheric Administration (NOAA). NOAA funding ran low during the first Bush administration, and NASA again entered the picture. By 1995, images with high resolution were available for commercial uses ranging from land management to insurance claims adjustment. Landsat 7

was launched in 1999. Commercial satellites followed. The IKONOS and the French Système Pour l'Observation de la Terre (SPOT) systems have resolutions closer to claimed American military capabilities. Google Earth uses satellite images to provide incredibly detailed views of Earth's human infrastructure.

As for U.S. assets, after the turn of the century, only Landsat 5 and 7 remained available. In August, 2007, Landsat 5 unexpectedly tumbled out of its working orbit. Several days later, that satellite was recertified for continued operations; some believed Landsat 5 had been hit by debris from the Perseid meteor shower. This anomalous orbit incident, however, illustrated another aspect of using the resources of space: the expanding danger of micrometeoroid and orbital debris (MMOD). Both LEO and GEO have become filled with operational satellites, space junk, spent booster parts, and other debris. Quite often the International Space Station (ISS) has to execute collision-avoidance maneuvers to miss orbital debris.

In 2009, an investigation of joint management NASA and the Department of the Interior U.S. Geological Survey indicated that Landsat was not meeting requirements of the 1992 Land Remote Sensing Policy Act. This investigation called for greater thermal imaging capability and urged an expanded Landsat Data Continuity Mission to maintain Landsat legacy data.

Astronomical Satellites

Atmospheric density fluctuations cause starlight to twinkle. Without adaptive optics built into land-based telescope facilities, optical images smear out and obscure detail. Placing the Hubble Space Telescope above the atmosphere in LEO (in 1990) enabled astronomers to begin resolving individual stars and distant galaxies much farther away than ever before. This provided a better measurement of the size of the observable universe and a more accurate value for the rate at which the universe is expanding, the so-called Hubble constant. Other astronomical satellites have detected radiation that is partially or completely blocked by Earth's atmosphere: infrared, ultraviolet, X-ray, and gamma-ray radiation. The Cosmic Background Explorer (COBE) measured diffuse infrared and microwave radiation thought to be remnants of the big bang and revealed tiny fluctuations that may have led to galaxy formation.

Vela satellites, launched in 1969 to monitor the Nuclear Test Ban Treaty, discovered unexpected celestial gamma-ray emissions. The utilization of Earth-orbiting and solar-orbiting positions for astrophysical studies of the cosmos at wavelengths not available to Earth-based observatories was quickly realized by such early spacecraft as the Orbiting Astronomical Observatories, the Orbiting Solar Observatories, and the High Energy Astronomical Observatories. The aforementioned Hubble Space Telescope became but one of a collection of Great Observatories that NASA launched into space. Others were the Compton Gamma Ray Observatory, the Chandra X-Ray Observatory, and the Spitzer Space Telescope. The latter was an infrared observatory. These Great Observatories permitted coordinated studies in several ranges of the electromagnetic spectrum, greatly expanding the understanding of high-energy astrophysics and cosmological issues.

NASA and other international space agencies also developed smaller space-based observatories designed for more specific investigations. Fermi and Swift extended gamma-ray studies by Compton and some Russian spacecraft. The French launched the Convection Rotation and Planetary Transits (COROT) telescope to look for transits of extrasolar planets across their star. NASA's Kepler spacecraft greatly exceeded COROT in capability and began looking for Earth-class planets in extrasolar systems in 2009.

Manufacturing in Microgravity

Any object in a circular orbit about Earth is in a state of free fall, having just enough speed (hence the right total mechanical energy) to fall around Earth instead of getting radially closer to its surface. This condition is weightlessness, a state wherein gravitational influence is balanced by centripetal motion. This description applies equally to elliptical orbits in which the orbiting object's speed varies as it undergoes periodic orbital motion. Effects such as the gravitational attraction of other bodies on an object may give that object a weight many orders of magnitude smaller than its normal "Earth" weight, a situation referred to as "microgravity."

When crystals are formed out of solution on Earth, they often develop imperfections because of convective flow within the solution. More nearly perfect crystals can be formed in microgravity,

because there are no gravitationally induced convection currents. Microgravity materials processing has proven to be useful, but it has yet to become cost-effective. As of 2009, it cost roughly $22,000 per kilogram to deliver a payload to orbit.

Research opportunities on the ISS in 2009 began to expand greatly under a plan to operate ISS as a national laboratory with international partners. As a result of ISS research, a salmonella vaccine developed in space was put into clinical trials on Earth. Other pharmaceutical projects on ISS held the potential for billions of dollars in profits in addition to lessening human suffering.

Solar Satellite Power Stations

Some have proposed using solar satellite power stations (SSPS's) to generate electrical energy. Ideas such as these go back as far as the late 1960's. A large SSPS in geostationary orbit might require 50 square kilometers or more of solar collectors. Electricity from those solar collectors could be converted into microwaves and be beamed down to a ground-based antenna array, where it could be converted into normal alternating electric current. In order

A rendering of an orbiting Block II-F (GPS) *satellite.* (NASA)

to maintain a safe microwave beam intensity, the antenna array would need to cover many square kilometers. Some have suggested that one or two hundred of these stations could supply all electrical needs of the United States.

The idea has certain attractions, especially if the receiving arrays could be situated in unpopulated regions. Solar power would generate no carbon dioxide emissions to aggravate global warming. On the other hand, there would be huge amounts of mining and manufacturing wastes associated with acquiring materials for constructing the receiving arrays and satellites. Lifting the satellite materials into orbit might require 30,000 to 60,000 space shuttle-class launches, which, beyond the idea's impracticality, would be an environmental disaster in and of itself. This idea remains popular among certain commercial space and public space advocacy groups but has generated little government support.

Mining the Moon and Mars

In 1969, Gerard K. O'Neill of Princeton University set up his freshman physics course as a seminar geared toward exploring whether a planetary surface was really the right place for an expanding technological civilization; the students returned a negative answer. However, consensus grew that colonies in space were feasible and could provide access to abundant energy, raw materials, freedom, and frontiers beyond Earth. O'Neill's disciples and successors have a remarkable idealism and a zeal about humankind's place in space. They organized as the Space Studies Institute (SSI) and the Space Frontier Foundation. Other advocacy groups arose, such as the 15 Society, named after a concept to place a huge human space colony at a specific Lagrange point in the Earth-Moon system.

Using solar energy and appropriate industrial chemical processes, extracting oxygen, silicon, iron, calcium, aluminum, magnesium, and titanium from lunar rocks and soil should be possible. Oxygen and powdered aluminum could be used as rocket fuel. Mass drivers, devices designed with tracks and sequentially activated magnetic coils to propel buckets of material to launch speeds, could launch supplies from the lunar surface. Space tugs could catch these supplies and transport them to a space colony. It would cost much less energy to bring material from the Moon to build an SSPS than it would to provide it from Earth. Even so, it is doubtful that the SSPS would pay for itself unless the space colony were already in place.

The Martian surface or perhaps Phobos, one of Mars's two small irregular moons, could become a spacecraft fueling station. Water could be mined from polar ice or from permafrost and be converted into high-grade rocket fuel based on hydrogen and oxygen. Carbon dioxide from the Martian atmosphere could be processed into a rocket-fuel combination of oxygen and carbon monoxide. The ability to refuel would make access to Mars and the asteroid belt easier. Aggressive exploration and exploitation of Mars have been advocated by Robert Zubrin and the Mars Society. Mars remains a long-range, albeit unfunded, goal of NASA manned spaceflight.

In the aftermath of the *Columbia* accident in 2003, the second Bush administration advanced the Vision for Space Exploration with the motto: "The Moon, Mars, and Beyond." The primary charge to NASA was to return to the Moon to stay, with initial lunar operations to begin by 2020. A goal of steadily building up a lunar base at the Moon's south pole, using as many in situ resources as possible, became NASA's Project Constellation. Other nations, including China, Russia, India, and Japan, developed interests in exploring lunar space as well. An implied Chinese manned spaceflight goal was to reach the Moon before NASA's return. Apollo 17 moonwalker Harrison Schmitt developed an economically sustainable plan to mine lunar soil for helium 3 to be used on Earth in fusion-based power generation systems. As of 2010, American plans for a return to the Moon were under review.

Mining Asteroids

Some asteroids are excellent sources of nickel and iron. Others contain a great deal of carbon and water. There are an estimated two thousand asteroids 1 kilometer in diameter or larger that cross Earth's orbit. These asteroids are more accessible than those within the main asteroid belt. It is at least theoretically possible to adjust the orbits of smaller asteroids using mass drivers or gravity tractors, but it might take years or decades to achieve the desired orbit. It is believed that a single nickel-iron asteroid 1 kilometer in diameter would contain nearly seven times the estimated earthly nickel reserves.

In 2009, British scientists presented a design for a gravity tractor that would fly close to an asteroid surface and, through gravitational influence alone, over perhaps fifteen years, make changes in the orbital path of such a body. If a near-Earth object or small asteroid were on a collision course with Earth, such a spacecraft placed close to its surface could avert a deadly global catastrophe. A 9-metric-ton gravity tractor, however, could not be used to bring a resource-laden asteroid into Earth proximity for convenient mining operations.

Charles W. Rogers and David G. Fisher

Further Reading

Clarke, Arthur C. *The Snows of Olympus: A Garden on Mars*. London: Victor Gollancz, 1994.

Davidson, Frank Paul, Katinka I. Csigi, and Peter E. Glaser. *Solar Power Satellites: A Space Energy System for Earth*. New York: Wiley & Sons, 1998.

Elbert, Bruce R. *Introduction to Satellite Communication*. 3d ed. New York: Artech House, 2008.

Fogg, Martyn J. *Terraforming: Engineering Planetary Environments*. Warrendale, Pa.: Society of Automotive Engineers, 1995.

Handberg, Roger. *International Space Commerce: Building from Scratch*. Gainesville: University Press of Florida, 2006.

Harris, Philip Robert. *Space Enterprise: Living and Working Offworld in the Twenty-first Century*. New York: Praxis, 2009.

Johnson, Richard D., and Charles Holbrow, eds. *Space Settlements: A Design Study*. Washington, D.C.: National Aeronautics and Space Administration, 1977.

Karl, John. *Celestial Navigation in the GPS Age*. New York: Paradise Cay, 2007.

Kendall, Henry W., and Steven J. Nadis, eds. *Energy Strategies—Toward a Solar Future: A Report of the Union* of Concerned Scientists. Cambridge, Mass.: Ballinger, 1980.

Lewis, John S. *Mining the Sky: Untold Riches from the Asteroids, Comets, and Planets*. Reading, Mass.: Addison-Wesley, 1996.

Olla, Phillip, ed. *Commerce in Space: Infrastructures, Technologies, and Applications*. Hershey, Pa.: Information Science Reference, 2008.

Olsen, R. C. *Remote Sensing from Air and Space*. New York: SPIE Press, 2007.

Pop, Virgiliu. *Who Owns the Moon? Extraterrestrial Aspects of Land and Mineral Resources Ownership*. Dordrecht, the Netherlands: Springer, 2009.

Ride, Sally K. *Mission, Planet Earth: Our World and Its Climate—And How Humans Are Changing Them*. New York: Flash Point, 2009.

Robinson, Ian S. *Measuring the Oceans from Space: The Principles and Methods of Satellite Oceanography*. New York: Springer, 2004.

Schmitt, Harrison. *Return to the Moon: A Practical Plan for Going Back to Stay*. New York: Springer, 2005.

TIDES AND WAVES

Category: Weather, Nature, and Environment.

Approximately 70 percent of the Earth's surface is covered with water, most of which is in a constant state of motion. The causes of this motion include the gravitational pull of celestial bodies in space, like the sun and moon; the rotation and shape of the Earth; and the influence of natural phenomena, like wind and earthquakes. Mathematicians have long studied tides and waves, following in the path of ancient scholars and others who sought to understand these phenomena for many spiritual and practical reasons, such as sailing. In the twenty-first century, people still travel both above and below the surface of the oceans for research, commerce, and pleasure, and there are many problems old and new to be explored. Some interesting mathematical investigations related to tides and waves at the start of the twenty-first century include three-dimensional modeling of extreme waves (also called "rogue waves"), such as those observed during the 2004 Indian Ocean tsunami and the Hurricane Katrina storm surges in 2005. Mathematicians, scientists, and engineers have also explored methods and developed technology to harness tide and wave power as an alternative energy source, including methods that actually create waves in addition to using naturally-occurring ones.

Overview

Water in Earth's oceans moves in a variety of ways, including many scales of currents, tides, and waves. Mathematicians and scholars from ancient times up through the Renaissance observed, identified, and quantified tidal patterns. The term "tides" generally refers to the overall cyclic rising and falling of ocean levels with respect to land—though tides have been observed in large lakes, the atmosphere, and Earth's crust, resulting largely from the same forces that produce ocean tides.

The daily tide cycles are caused by the moon's gravity, which makes the oceans bulge in the direction of the moon. A corresponding rise occurs on the opposite side of the Earth at the same time, because the moon is also pulling on the Earth itself. Most regions on Earth have two high tides and two low tides every day, known as "semidiurnal tides," which result from the daily rotation of the Earth relative to the moon. Since the angle of the moon's orbital plane also affects gravitational pull on Earth's curved surface, some regions have only one cycle of high and low, known as "diurnal tides." The

Officers of the National Oceanic & Atmospheric Administration Corps photographed the devastation caused in New Orleans by the 2005 Hurricane Katrina's storm surges. (NOAA Aviation Weather Center)

height of tides varies according to many variables, including coastline shape; water depth ("bathymetry"); latitude; and the position of the sun, which also exerts gravitational force. "Spring" tides, not named for the season, are extremely high and low tides that occur during full and new moons when the sun and moon are in a straight line with the Earth, and their gravitational effects are additive. A proxigean spring tide occurs roughly once every 1.5 years when the moon is at its proxigee (closest distance to Earth) and positioned between the sun and the Earth. Neap tides minimize the difference between high and low tides. They occur during the moon's quarter phases when the sun's gravitational pull is acting at right angles to the moon's pull with respect to the Earth.

A few of the many contributors to the theory and mathematical description of tides include Galileo Galilei, René Descartes, Johannes Kepler, Daniel Bernoulli, Leonhard Euler, Pierre Laplace, George Darwin, and Horace Lamb. Some mathematicians, like Colin Maclaurin and George Airy, won scientific prizes for their research. Work by mathematician William Thomson (Lord Kelvin) on harmonic analysis of tides led to the construction of tide-predicting machines.

Waves

There are many mathematical approaches to the study of waves in the twenty-first century, and some mathematicians center their research around this topic. In contrast to tides, a wave is a more localized disturbance of water in the form of a propagating ridge or swell that occurs on the surface of a body of water. Despite the fact that surface waves appear to be moving when observed, they do not move water particles horizontally along the entire path of the wave. Rather, they combine limited longitudinal or horizontal motions with transverse or vertical motions. Water particles in a wave oscillate in localized, circular patterns as the energy propagates through the liquid, with a radius that decreases as the water depth or distance from the crest of the wave increases. Wind is a primary cause of surface waves, because of frictional drag between air and water particles. Larger waves, like tsunamis, result from underwater Earth movements, such as earthquakes and landslides.

The Navier–Stokes equations, named for Claude-Louis Navier and George Stokes, are partial differential equations that describe fluid motion and are widely used in the study of tides and waves. Solutions to these equations are often found and verified using numerical methods. The Coriolis–Stokes force, named for George Stokes and Gustave Coriolis, mathematically describes force in a rotating fluid, such as the small rotations in surface waves. A few examples of individuals with diverse approaches who have won prizes in this area include Joseph Keller, who has researched many forms and properties of waves, including geometrical diffraction and propagation; Michael Lighthill and Thomas Benjamin, who jointly posed the Benjamin–Lighthill conjecture regarding nonlinear steady water waves, which continues to spur research in both theoretical and applied mathematics; and Sijue Wu, who has researched the well-posedness of the fully two- and three-dimensional nonlinear wave problem in various function spaces, using techniques like harmonic analysis. In other theoretical and applied areas, some techniques from dynamical systems theory, statistical analysis, and data assimilation, which combines data and partial differential equations, have been useful for formulating and solving wave problems.

Sarah J. Greenwald
Jill E. Thomley

Further Reading

Cartwright, David. *Tides: A Scientific History*. Cambridge, England: Cambridge University Press, 2001.

Johnson, R. I. *A Modern Introduction to the Mathematical Theory of Water Waves*. Cambridge, England: Cambridge University Press, 1997.

Joint Policy Board for Mathematics. "Mathematics Awareness Month April 2001: Mathematics and the Ocean." http://mathaware.org/mam/01.

THE TUNGUSKA EVENT

Category: Impact Events

The Tunguska event took place on June 30, 1908, in Tunguska, Siberia. A meteorite or comet is believed to have exploded in the air, releasing energy equivalent to at least 10 to 20 megatons of TNT and leaving two people dead, several nomad camps destroyed, more than 1,000 reindeer killed, and 811 square miles (2,100 square kilometers) of forest flattened. The event is often referred to as an impact event.

Overview

Early on the morning of June 30, 1908, witnesses along a 621-mile (1,000-kilometer) path saw a fireball streak across the sky from the east-southeast. It was as bright as the Sun and cast its own set of shadows in the early morning light. The object exploded at 7:14 A.M., local time. Based upon seismic and barographic records, and upon the destruction caused, the explosion released energy equivalent to that of ten to twenty megatons of TNT, making it the most devastating cosmic event on Earth during historical times. Depending upon the altitude of the explosion and the composition of the object, the energy released may have been as high as 50 megatons.

Had the explosion occurred over New York City, fatalities would have been in the millions. As it was, the object exploded over a sparsely inhabited forest in Siberia, roughly 43.5 miles (70 kilometers) north of Vanavara, a small village on the Stony Tunguska River. The region is one of primeval forests and bogs inhabited by nomads who tend large herds of reindeer. Near the epicenter (ground zero), trees burst into flame. Farther out, a great shock wave felled trees over an 811-square-mile area, pointing them radially outward, bottoms toward, and tops away from the epicenter. Right at the epicenter where the force of the blast wave was directly downward, a bizarre grove remained. Trees were left standing upright, but they were stripped of all their branches, like telephone poles.

An eyewitness in Vanavara said the sky was split apart by fire and that it was briefly hotter than he could endure. Because it was just after the summer solstice, the Sun remained above the horizon twenty-four hours a day north of the Arctic Circle. Dust, lofted high into the stratosphere, scatted so much sunlight back to the ground that even south of the Arctic Circle, in northern Europe and Asia, nights were not really dark for three days. People were amazed that they could read, or even take photographs, in the middle of the night. At least 1,000 reindeer were killed, and several nomad camps were blown away or incinerated. Some nomads were knocked unconscious, but remarkably, there are only 2 known human fatalities. An old man named Vasiliy was thrown 39 feet (12 meters) through the air into a tree. He soon died of his injuries. An elderly hunter named Lyuburman died of shock.

Scientists supposed that the seismic waves had been caused by an earthquake, but no scientists went immediately to investigate because of the remoteness of the site. It was not until 1927 that Leonid Kulik, the founder of meteorite science in Russia, reached the site after spending many days plunging through trackless bogs on horseback. Expecting to find a huge crater and a valuable nickel-iron mountain, Kulik and his assistant were amazed to find only a shattered forest stretching from horizon to horizon.

Careful research has since shown that the Tunguska object shattered about 5.3 miles (8.5 kilometers) above the ground. If it were a small comet, it must have been inactive, for there is no credible evidence of a tail. It must have been at least 328 feet (100 meters) in diameter and had an asteroidal core, because microscopic metallic particles were recovered that are more closely associated with asteroids than with comets. Russian scientists favor this hypothesis. The object's trajectory and timing are consistent with it being a fragment of Comet Encke. Western scientists favor the possibility that it was a small, dark, rocky asteroid, perhaps 197 feet (60 meters) in diameter.

When a solid object of this size plunges into the atmosphere, it piles up air in front of it until the air acts like a solid wall. The object shatters, its kinetic energy is converted to heat, and the object vaporizes explosively. Microscopic globules form as the vapor condenses. Such globules have been recovered from peat bogs and tree resin at the site, as well as from ice layers in remote Antarctica. The cosmic dust cloud truly spread worldwide. These globules have more of the elements nickel and iridium than normal Earth rocks do—clear signatures of their cosmic origins.

Charles W. Rogers

Further Reading

Chaikin, Andrew. "Target: Tunguska." *Sky and Telescope*, January, 1984, 18-21.

Fernie, J. Donald. "The Tunguska Event." *American Scientist*, September-October, 1993, 412-415.

Gallant, Roy A. "Journey to Tunguska." *Sky and Telescope*, June, 1994, 38-43.

YUCATÁN CRATER IMPACT

Category: Impact Events

About 65 million years ago, an impact event occurred in the Yucatán Peninsula which released energy equivalent to at least 100 million megatons of TNT, instantly destroyed most life within 621-mile radius, and caused worldwide climate changes resulting in the extinction of up to 85 percent of species then living

Overview

A team of scientists led by Luis and Walter Alvarez, father and son, were studying the thin clay layer that lies between the rocks of the Cretaceous geological period and the rocks of the following Tertiary period. This boundary is designated the K/T boundary. (By convention, Cretaceous is abbreviated *K*. The letter *C* is used for the earlier Cambrian period.) Knowing that the element iridium is more abundant in meteorites than in earth rocks, and supposing that small meteorites fall at a more or less constant rate, they supposed that the amount of iridium in the clay would be a clue to how long it took to form the clay layer. To their great surprise, they discovered that the iridium concentration in the clay was three hundred times that of the rocks above and below it. In 1980, they startled the world with this result and with their theory of what had ended the reign of the dinosaurs 65 million years ago.

According to their theory, now widely accepted, a rocky asteroid 6.2 miles (10 kilometers) or more in diameter hurtled toward Earth at tens of miles per second. Plunging through the atmosphere in a few seconds, its energy of motion was converted into heat as it struck the ground, vaporizing itself along with a great deal of the target rock. The resulting explosion lofted 100 million megatons of dust and rock vapor into the air, much of it out into space. It also produced an earthquake 30,000 times more energetic than the San Francisco earthquake of 1906.

There is a huge crater about 112 miles (180 kilometers) across at Chicxulub, Yucatán. It is 65 million years old and is thought to be the impact site of the Alvarez asteroid. Fittingly, Chicxulub (pronounced CHEEK-shoe-lube) means "tail of the devil." Today, the crater is completely covered with surface rock. Further evidence of an impact is that all around the Gulf of Mexico there is a 65-million-year-old layer of tsunami-wave rubble 33 feet (10 meters) thick, including large boulders washed far inland. Shock-fractured crystals found in the K/T boundary

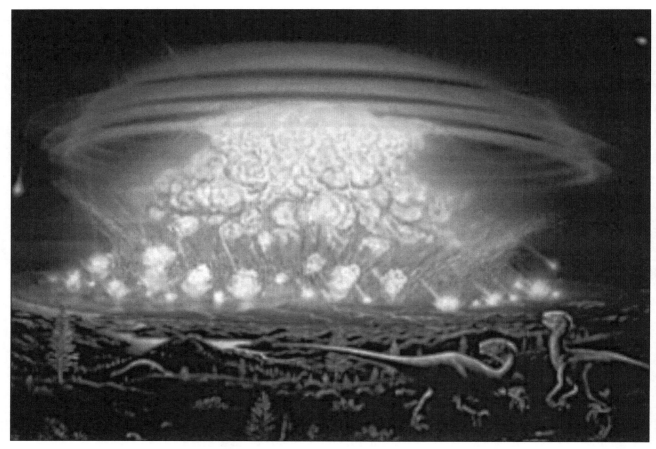

An artist's rendering of the meteorite that struck Earth 65 million years ago, wiping out the dinosaurs.(NASACORE/LorainValley JVS)

layer are another key piece of evidence. While a large impact can form these crystals, volcanic activity cannot.

Shock and heat from the impact killed nearly everything above ground within 621 miles (1,000 kilometers). The vapor that was lofted into space cooled and condensed into rocky globules that reheated as they plunged back into the atmosphere all around the world. Their heat started forest fires worldwide. The amount of soot found in the worldwide K/T boundary layer shows that much of Earth's total biomass burned. Smoke from these fires combined with dust lofted into the stratosphere by the impact formed a worldwide pall that blocked sunlight for months, causing Earth to cool about 40 degrees Fahrenheit and photosynthesis to cease. This has been called "impact winter."

Heat from the fireball caused nitrogen and oxygen in the atmosphere to combine to form nitric oxide, which was lofted into the stratosphere, where it destroyed the ozone layer. Less than 2 percent of Earth's surface is covered with layers of limestone and evaporite 1.2 to 1.9 miles (2 to 3 kilometers) thick, but the Yucatán Peninsula is such a place. Vaporizing these deposits released huge amounts of sulfur dioxide and carbon dioxide. Nitric oxide and sulfur dioxide combined with water vapor in the air to form acid rain. There may not have been enough acid rain worldwide to be a serious problem by itself, but it did add to the environmental insult. As the dust cleared, "impact winter" turned to "impact summer," and the climate warmed about 40 degrees Fahrenheit above normal for thousands of years. These elevated temperatures were possibly due to a greenhouse effect caused by the extra carbon dioxide and water vapor in the atmosphere.

Which species became extinct and exactly when that happened remains somewhat controversial; however, the most complete studies support the hypothesis that the dinosaurs died because of the climate-changing effects of an asteroid impact. The general pattern is that species such as dinosaurs, whose food chain depended upon living plant material, became extinct. Species whose food chain depended upon organic detritus left in logs, soil, or water survived and eventually expanded into niches previously dominated by extinct species. Apparently, mammals survived on insects, arthropods, and worms until the sun began to shine again and plants to grow again.

Charles W. Rogers

Further Reading

Alvarez, Luis W. "Mass Extinctions Caused by Large Bolide Impacts." *Physics Today*, July, 1987, 24-33.

Beatty, J. Kelly. "Killer Crater in the Yucatán?" *Sky and Telescope*, July, 1991, 38-40.

Raup, David M. *The Nemesis Affair: A Story of the Death of Dinosaurs and the Ways of Science*. New York: W. W. Norton, 1986.

Verschuur, Gerrit L. *Impact! The Threat of Comets and Asteroids*. New York: Oxford University Press, 1996.

INDEX